Fish Biotechnology and Genetics

The Author

Dr. N. L. Nyan, expert in Fish Biotechnology and genetics, he served Department of Fisheries and Aquatic Sciences, University of Cape Coast, Cape Coast, Ghana. Dr. Nyan is Ph.D. in Fish Biotechnology, possesses a career of leadership in Fish Biotechnology Development in higher education.

Fish Biotechnology and Genetics

Dr. N. L. Nyan

2025

Agri Horti Press

New Delhi-110 084

ISBN 978-93-87642-69-0 (Hardbound)

Published By
Agri Horti Press
New Delhi
E-mail: agrihortipress@gmail.com
Printed at : **Replika Press Pvt. Ltd.**

PREFACE

Aquaculture Fisheries Biotechnology and Genetics is an ambitious title with broad coverage of techniques of fishes and their culture. It has been designed for university students scholars and teachers dealing with Genetics, molecular biology, genetic engineering, recombinant DNA technology. Fish and fisheries and biochemistry. The genetic improvement of fish for aquaculture and related fisheries is a field or research that has seen immense advances over the years. The molecular genetics programs are focused on developing highly specific and highly sensitive technologies to assist aquaculture industry. Current research across the globe is at advancing basic understanding of biological and genetic processes in order to guide the development of appropriate policies and protocols governing the application of biotechnology in fisheries an aquaculture. Fisheries and aquaculture emerged as the important commercial activity from its traditional role as subsistence supplementary activity.

With increased demand for aquacultured foods has come a need for more efficient production systems. Major improvements have been achieved through enhanced husbandry procedures, improved nutrient, enhanced disease diagnosis and therapies and the application of genetics to production traits. The present book "Fish Biotechnology and Genetics" describes introduction, fish anatomy, biotechnology in fisheries development, genetics in fisheries, genetic aquaculture resource, fish farming and aquaculture, conservation and utilization of genetic resources , biotechnology in fisheries and aquaculture . This book is extremely useful for students, teachers and researchers of fisheries science and all those working on fisheries and aquaculture management.

N. L. Nyan

Contents

Preface *v*

1. Introduction 1

2. Fish Anatomy 38

3. Biotechnology in Fisheries Development 69

4. Genetics in Fisheries 120

5. Genetic Aquaculture Resources 160

6. Fish Farming and Aquaculture 208

7. Conservation and Utilization of Genetic Resources 243

8. Biotechnology in Fisheries and Aquaculture 256

Bibliography *277*

Index *279*

1

Introduction

Status and Challenges of Fisheries

The fisheries administrators have viewed MCS as little more than the policing of the various maritime zones controlled by the State. This chapter attempts to provide a broader view of MCS - a view of MCS as the vital executive arm of fisheries management.

The rapid depletion of key fish stocks in the 1980s and 1990s has made it imperative that governments achieve greater control over fishing activities.

At the international level, a number of new agreements have created a stronger legal basis on which to develop greater control. At the same time, new technological developments have facilitated the remote monitoring of fishing vessels and the collection of fisheries data.

The 1982 United Nations Convention on the Law of the Sea (UNCLOS), which entered into force in 1994, forms the backbone of the international legal framework for fisheries management. It sets out the rights and duties of coastal, port and flag States in respect of each of the principal maritime zones recognized by international law, namely the territorial sea, the exclusive economic zone and the high seas. It also deals with a range of other important issues that are related, including the legal regimes applicable to internal waters, archipelagic waters, the contiguous zone, the continental shelf, and the right of innocent passage and passage through international straits.

The provisions of UNCLOS relating to fisheries, though widely accepted even before the Convention entered into force, did not prevent the depletion of several valuable fish stocks. For this reason, the 1992 United Nations Conference on Environment and Development (UNCED) called urgently for the development of further instruments that would be necessary to re-establish and maintain sustainable fisheries worldwide.

One of these new instruments is the Agreement for the Implementation of the Provisions of the United Nations Convention on the Law of the Sea of 10 December 1982 Relating to the Conservation and Management of Straddling Fish Stocks and Highly Migratory Fish Stocks. The UN Fish Stocks Agreement sets forth a broad range of obligations designed to create greater control over fisheries for certain valuable stocks, including the strengthening of MCS capabilities.

The pertinent aspects of several other new instruments developed under the auspices of FAO, including:

- The 1993 Agreement to Promote Compliance with International Conservation and Management Measures by Fishing Vessels on the High Seas;
- The 1995 Code of Conduct for Responsible Fisheries (CCRF);
- Four International Plans of Action dealing with various aspects of fisheries management, and particularly the International Plan of Action to Prevent, Deter and Eliminate Illegal, Unreported and Unregulated Fishing;
- Guidelines for the marking and identification of fishing vessels.

Ideally, each State would implement the 1982 Convention and the more recent agreements through the development of a national "oceans policy." Such a policy would establish government priorities and the strategy for the conservation and sustainable use of all marine resources within the maritime zones over which the State exercises sovereignty or sovereign rights, as well as efforts towards cooperative management of fisheries that occur outside these zones.

From this oceans policy would flow the integrated oceans planning and management framework under which fisheries management plans would be developed. Most States see this as a long-term development initiative.

To shorten the process, they have chosen to develop oceans policy and fisheries management strategies (including MCS strategies) simultaneously. Consequently, although States recognize that fisheries management must be integrated into an overall oceans policy when it is ultimately established, the MCS systems required to implement fisheries management plans are being developed in the interim to address the more immediate need to protect fish stocks and their habitats. This strategy is commendable. A "precautionary approach" recognizes that a first requirement for fisheries resource conservation is to prevent further degradation of the resource base.

The degree to which a government becomes involved in the fishing industry will have an impact on fisheries management and the resultant MCS activities.

For example, a government can:

- Assume a controlling role, where it actually runs the industry, impacts the potential income of fishers, and micro-regulates the harvesting sector; or

- Maintain a less intrusive or co-management role, whereby the fishers and the fishing industry are encouraged to accept their resource responsibilities and roles within the framework of general government conservation principles and legislation.

Negative results of centralised, micro-management control mechanisms have become evident in both industrialised and developing States. Consequently, there is an emerging trend towards the second, *participatory co-management approach.* Fishers and the fishing industry want, and in fact are demanding, a more active role in management planning and implementation. Central governments are responding by devolving authority to smaller units of government (provinces, districts and municipalities) and by fostering community-based management, stakeholder involvement, and the acceptance of responsibility for the care, conservation and protection of their local marine resources by the fishers and industry. In a growing number of cases, such as the Canadian experience described in Profile 1, the private sector is also becoming involved in MCS activities.

Misperceptions of MCS

Misperceptions surrounding MCS activities continue to impact how fishers and even fishery managers view the process of MCS and enforcement. A common misperception is that all fisheries problems stem either from a failure to control illegal foreign fishing, or from the fishers themselves. While foreign fishing fleets have had documented impacts on fisheries conservation efforts, the greater impact on fisheries often stems from the domestic fishing industry in the coastal and nearshore fishing zones. On the issue of fishers, fisheries administrators must remember that most are hard working individuals often working in a hazardous environment:

While sometimes libellously assumed by the ill-informed to be crooks, [fishers] are perhaps best described as being as honest as the next man, but hard, individualistic businessmen running very competitive and often highly capitalized operations. It is worth remembering that they do so in the face of a largely unforgiving sea that creates a working environment which... [has one of the worst industrial accident rates in the world], and [that]... they operate increasingly in an economic climate of ever increasing overheads countered only by the proceeds of catches which [are] subjected to [ever greater] quota restriction.

All of this is done in the knowledge that the success of their venture and the livelihood of their crews depends entirely on their individual skill, effort and initiative. Given these pressures, it is perhaps not surprising that such independent minds do not always take kindly to bureaucratic controls, especially if these appear to them to have little practical purpose. Another erroneous perception is that MCS is exclusively concerned with enforcement - thus ignoring the other two components of monitoring (data collection) and control (legislation, licensing, and controls on gear, season, areas, etc.). In focussing only on the "surveillance/enforcement" or deterrent aspects of MCS, fisheries administrators and supporting agencies cannot harness the full utility of MCS as the vital executive arm of fisheries management.

Civilian versus Military Involvement in MCS

The expense of MCS activities is often a primary concern of any government designing and implementing an MCS system. Cost-effectiveness and efficiency is important if MCS operations are to be successful. A civilian approach to deterrent fisheries enforcement has proven in many cases to be the most cost-effective and responsive to fisheries priorities. Use of civilian assets also minimizes the political sensitivity of international fisheries incidents by avoiding the use of military equipment and personnel. Those fisheries administrators who must rely on the use of military resources to carry out MCS activities may find that military agencies often accord low priority to that task.

Moreover, military involvement, except in a support role, is usually not cost-effective. Military aircraft and vessels are more expensive to build and operate than equivalent civilian equipment. Savings accrue from the use of a civilian vessel with fewer crew, and lower operating costs. For many governments, however, the military can play a significant supporting role in a strong MCS system. The key for such governments is to establish an inter-agency mechanism that enables fisheries administrators to call upon their military counterparts as and when needed.

Fisheries as a Lead Ministry

Effectiveness of operations can be enhanced considerably if a single ministry is designated to take the lead role in MCS activities. This significantly reduces the lines of communications for the command and control of the monitoring and surveillance components of MCS activities, making them more efficient and responsive to management needs. As noted above, however, a number of different agencies may be called upon in a supporting role. In such situations, effective MCS requires a strong inter-agency control mechanism.

In Canada, the federal Department of Fisheries and Oceans (DFO) has been entrusted by the Parliament of Canada through the *Fisheries Act* and the *Coastal Fisheries Protection Act* to administer all laws relating to fisheries. The administration of federal fisheries laws has, by agreement, been delegated to some, but not all, provincial governments. DFO remains responsible for fisheries management in the tidal waters of the Pacific, Atlantic, Arctic, the inland waters of four Atlantic provinces and the salmon rivers in British Columbia. This includes management of Aboriginal, Recreational and Commercial fisheries within Canada's Exclusive Economic Zone (EEZ), in transboundary rivers, and for sedentary species on the continental shelf outside the Canadian EEZ.

Statistics:

- 58 400 commercial fishers (42 700 Atlantic, 8 700 Pacific and 7 000 inland)
- Commercial harvest > 1 000 000 mt
- Landed value approximately $1.9 billion (shellfish account for approximately 75% of the total landed value, with groundfish and pelagic fisheries making up the remainder)
- 5 million Canadians and 900 000 visitors/year for the recreational fishery
- Catch over 250 million fish in recreational fisheries, with more than half of the fish released under the catch and release regime
- Over 125 fisheries agreements are negotiated with Aboriginal groups in Canada on an annual basis to provide for aboriginal access to fisheries and an orderly management of their fishing activities.

Management Systems/Control Mechanisms

Various management schemes have been applied in different fisheries, but all are based on limited entry licensing, with vessel and gear restrictions. Other measures to limit catches include total allowable catches (*e.g.* groundfish), escapement targets (salmon), or recruitment strategies (*e.g.* lobster). Other management measures include limitations on fishing area, fishing season, gear (*e.g.* mesh size), incidental catch (bycatch) and minimum fish sizes. Rights-based systems - in the form of Enterprise Allocations (EA) and Individual Quotas (IQ) - have been introduced in some fisheries to allow fishers more efficiently to manage the capacity and effort for harvesting.

MCS Programme

The Department of Fisheries and Oceans (DFO) Conservation and Protection programme ensures compliance with the legislation, regulations and fishing plans. This requires an integrated MCS approach and deployment of some 600 Fishery Officers for air, sea and land patrols; independent/private sector observer coverage on fishing vessels; dockside monitoring of fish landings; and remote electronic monitoring of fishing vessel activity.

DFO operates a fleet of patrol vessels on each coast to enforce closed areas and boundary lines, and to conduct inspections at sea for compliance purposes. Contracted aircraft are used to monitor fishing fleets. Sea and aerial surveillance is also supplemented by the Department of National Defence (DND). VMS and air surveillance provide for more effective deployment of patrol vessels; with the latter also serving as a visible deterrence. Canada deploys private sector, contracted observers, without enforcement powers, on all foreign vessels fishing in Canadian waters and on some Canadian vessels to gather scientific information and provide on-site monitoring of compliance.

They are trained to detect and report infractions such as dumping/discarding, fishing in closed areas, catch misreporting, retention of prohibited catch and the use of illegal gear. The level of observer coverage in domestic fisheries varies depending on conservation risks and management priorities. Contracted dockside monitors/observers verify the quantity and species/product form of fish landed for scientific, quota monitoring and compliance reasons. This data is cross-checked by random inspections by Fishery Officers at landing sites. At-sea observer costs are shared by the Department and the fishing industry. Dockside monitoring costs are entirely paid by the fishing industry. In 1999, DFOs expenditures on fisheries enforcement amounted to CDN $70 million (including DFOs share of at-sea observer costs).

The Conservation and Protection programme has been significantly re-oriented in recent years. The mix of enforcement resources has been altered to better respond to changing programme requirements. For example, a number of larger patrol vessels have been replaced with smaller programme boats that can be operated more efficiently by the Fishery Officers themselves. Savings from vessel reductions have been partially re-invested in new equipment and surveillance technologies. Significant investments have also been made in the creation of new enforcement data systems and the integration of existing systems, with the goal of providing Fishery Officers and managers with more accurate and timely information that will strengthen the Department's enforcement capabilities. These efforts at improving data integration and analysis will continue to be a priority for the immediate future.

The Conservation and Protection programme is closely integrated with DFOs overall Fisheries Management Programme. Input and advice from fisheries enforcement officials is an important consideration in the development of Integrated Fisheries Management Plans (IFMPs). In future these plans will include specific conservation objectives as identified by

DFO scientists in consultation with fishers and other technical experts. The new Objectives Based Fisheries Management strategy will require greater involvement by the fishing industry to design management measures that will minimize identified risks to conservation.

Conserve and Utilise Marine resources

Fisheries are critical to the development of a State's plan to conserve and utilise marine resources, as fish and their habitat are significant renewable resources in the territorial sea and exclusive economic zone. The goal of fisheries management, including MCS, is to maximise the economic opportunities and benefits from the State's waters within sustainable harvesting limits. Fisheries MCS needs to be defined in light of this goal. An MCS Conference of Experts organized by FAO in 1981 developed a definition of MCS that is commonly accepted by fisheries personnel:

- Monitoring - the continuous requirement for the measurement of fishing effort characteristics and resource yields;
- Control - the regulatory conditions under which the exploitation of the resource may be conducted; and
- Surveillance - the degree and types of observations required to maintain compliance with the regulatory controls imposed on fishing activities.

Simply stated, MCS is the mechanism for implementation of agreed policies, plans or strategies for oceans and fisheries management. MCS is an aspect of oceans and fisheries management that is often undervalued. In reality, it is key to the successful implementation of any planning strategy. The absence of MCS operations render a fisheries management scheme incomplete and ineffective. Since the 1981 MCS Conference, the definition of MCS has been enhanced to promote the concept that MCS covers more than just fisheries enforcement - it is an integral and key component for the implementation of fisheries management plans. It encompasses not only traditional enforcement activities but also the development and establishment of both data collection systems, the enactment of legislative instruments and the implementation of the management plan through participatory techniques and strategies.

A 1993 workshop in Ghana offered the following clarifications:

- *Monitoring* includes the collection, measurement and analysis of fishing activity including, but not limited to: catch, species composition, fishing effort, bycatch, discards, area of operations, etc. This information is primary data that fisheries managers use to arrive at management decisions. If this information is unavailable, inaccurate or incomplete, managers will be handicapped in developing and implementing management measures.

- *Control* involves the specification of the terms and conditions under which resources can be harvested. These specifications are normally contained in national fisheries legislation and other arrangements that might be nationally, subregionally, or regionally agreed. The legislation provides the basis for which fisheries management arrangements, via MCS, are implemented. For maximum effect, framework legislation should clearly state the management measures being implemented and define the requirements and prohibitions that will be enforced.
- *Surveillance* involves the regulation and supervision of fishing activity to ensure that national legislation and terms, conditions of access, and management measures are observed. *This activity is critical to ensure that resources are not over exploited, poaching is minimized and management arrangements are implemented.*

These wider definitions amplify the importance of all aspects of MCS.

Emerging Trends in MCS

Trends over in the past decade have significantly altered fisheries management in general and MCS strategies in particular. These have occurred in three areas - viz: *devolution of authority to lower levels of government; the encouragement of participatory management;* and *new technology.*

Devolution of Authority

Fisheries management can no longer be the sole responsibility of central governments. Rapid population growth, urbanisation and migration to coastal areas have created a greater need for local governmental units to address local fisheries management concerns. Growing literacy and educational levels, coupled with easy access to mass media, have helped ordinary citizens to acquire more knowledge of government processes and a desire for more transparency and a greater voice in decision-making. Central governments have responded by devolving more authority over fisheries management to provinces, states, districts, municipalities and communities.

This devolution of authority has vitally affected MCS. In the past, central governments controlled, or at least sought to control, all MCS activities on a national basis. Today, a more typical role for a central government is to set national policies and standards and to co-ordinate the implementation of these policies and standards. Responsibility for actually carrying out MCS activities is now typically shared by central governments with smaller governmental units. Central governments also monitor and evaluate local government performance to enhance future coordination and planning. MCS planning and operations still need this national coordination for consistency in implementation of management plans, and for conflict resolution. However, the overall trend is for central governments to devolve responsibility for MCS activities to lower levels of governmental authority.

Namibia, a State six times the size of the United Kingdom and with a population of 1.7 million, is on the West Coast of Africa just north of South Africa. It has a coastline of 700 nautical miles (1 500 km) and an EEZ of 275 000 km^2. Since independence in 1989, Namibia has developed a progressive fisheries administration and a thriving commercial capture fisheries that is increasingly Namibian.

The fishing industry is based on the high productivity of the up-welling Benguela Current system. Fisheries contributed more than US$50 million (10% Namibia's GDP) in 1999 and employed some 15 000 people. By law, all fish are landed in Namibia through its two ports, Walvis Bay and Lüderitz. Namibia is one of the few major fishing nations that earns more income and creates more jobs from the processing sector than from the catching of fish. This is partly due to policies such as the requirement for onshore processing of hake and other species. *Namibia is recovering all expenses related to its fisheries administration and MCS operations through the collection of revenue from the fishing industry.*

The fisheries are exclusively industrial and annual catches are stable or steadily increasing following the overexploitation before independence. Trends for the three major species (hake, horse mackerel and pilchard) are showing the same broadly consistent pattern of increases in biomass up to 1992, a subsequent decline and, recently, a general recovery. The total landed catch in 1999 was approximately 600 000 t.

The key fisheries are as follows (TAC for 2000 in brackets):

- *The Demersal Fisheries*: Hake (200 000 t), monk (16 000 t), kingklip and sole;
- *The Midwater Fishery*: Horse mackerel (410 000 t) pursued by both midwater trawlers and the purse seiners;
- The Purse Seine Fishery: Pilchard (15 000 t) and juvenile horse mackerel with purse seine nets;
- *The Deep-Water Fishery*: Orange roughy (2 400 t) and alfonsino;
- *The Large Pelagic Fishery*: Albacore, big-eye, yellow-fin, skipjack and swordfish;
- *The Rock Lobster Fishery (400 t)*: Small vessels using carrier vessels to bring the live lobster ashore daily;

- *The Crab Fishery (2 000 t)*: A small fishery that uses traps to catch deep-sea crab during the whole year.

Namibia uses a three-part access system to turn its policies of Namibianisation of the fisheries into practise:

i. Rights to exploit the fisheries are allocated;
ii. Quotas are allocated to rights holders; and
iii. Vessels are granted licences.

This allows Namibia to maintain control over the fishing companies, the vessels and the crews in order to assure that policy aims are being met through a differentiated quota fees system that favours Namibian participation, Namibian flag vessels, Namibian employment, and an empowerment aspect in relation to the social inequalities.

Fisheries management controls fall into two categories:

- Input controls that relate to fishing effort and gear, seasons and areas, with the key ones being the limitation of total fishing effort and seasons; and
- Output controls that set quotas and regulations, size limits, and other characteristics of the fish that may be landed. The main output control is by the establishment of annual TAC and quota allocations set on the best scientific advice available, modified by socio-economic factors. TACs are today established for 7 species (hake, horse mackerel, pilchard, orange roughy, monk, red crab and rock lobster).

The MCS system is an integrated system that has stations in the two Inspectorates in Walvis Bay and Lüderitz with each station tasked as appropriate for deploying fishery officers to air, sea or land operations, deploying fisheries observers on board fishing vessels, analysing past operations and outputs or planning future operations. There is a cross-verification of data where observer and inspector information is checked through port inspections and landings control. MCS activities related to air and sea operations are co-ordinated from the Inspectorate at Walvis Bay through a maritime and fisheries Operations Centre and an air base at Arandis.

The Ministry of Fisheries and Marine Resources operates two larger patrol vessels and one fixed wing aircraft in addition to maintaining 100% observer coverage of the fishing fleet (230 observers are employed in the largest observer programme south of equator). Observers and fisheries inspectors control gear and catches at sea. All catches are finally landed or transhipped in Namibia under the supervision and control of fisheries inspectors. VMS is still in a pilot stage and different systems are under evaluation.

Participatory Management

Similarly, governments at all levels are giving private groups and individuals more opportunity to participate in the management of fisheries that affect them. This trend towards more transparency and openness is changing all aspects of MCS, including the development of fisheries laws and regulations, the collection of fisheries data, etc. As already indicated, enhanced public participation in the fisheries management process brings a number of advantages.

Specifically, it fosters:

- Greater public understanding of the rationale behind MCS activities;
- A greater sense of partnership between the government and fishers and others whose activities are being regulated;
- A greater sense of ownership and acceptance by private groups and individuals in the decisions that are ultimately adopted;
- Voluntary compliance and the use of peer pressure to deter violations of fisheries rules;
- Greater availability of MCS funding and other assets for dealing with cases of non-compliance.

New Technology

Advances in technology, along with reductions in the cost of technology, have revolutionised MCS.

Examples include:

- New vessel monitoring systems (VMS) in use in Argentina, Australia, Canada, Europe, Malaysia, New Zealand, United States, and member States of the South Pacific Forum Fisheries Agency, with systems also being considered in Madagascar, Namibia, Indonesia, and Sri Lanka;
- Inter-linked coastal radar systems in use in Senegal; and
- A combined VMS and satellite imaging system in use in the Maldives.

Cheap cellular telephone and computer technology has also brought MCS into the computer age and has greatly improved the capacity of States to respond rapidly to situations. Over-the-horizon radar technology is evolving as well, but its cost has not yet fallen to a level that is appropriate for general fisheries management.

Currently the most popular VMS systems use INMARSAT C (and more recently D) and ARGOS. Others such as POLESTAR and EULTRACS are also making an appearance. FAO has prepared detailed guidelines as part of the *Code of Conduct for Responsible Fisheries* series for Fisheries

Administrators contemplating the introduction and use of such technology.

The use of this new technology for coastal fisheries, particularly VMS, may not yet be cost-effective. However, many States can still monitor large numbers of coastal vessels effectively through other means, *e.g.* observers, dockside monitoring, monitoring of industry landings and production records, etc.

Role of MCS in Fisheries Management

Fisheries management in its simplest terms comprise the following activities:

- *Data collection and analysis*: Data for management planning and operations from socio-economic studies, rural development studies, fisheries population studies, fisheries research cruises, licensing (national, provincial and district), catch and effort/logbooks, onboard observers (if established as a programme), dockside monitoring/landings, VMS, satellite imaging, inspections at sea and in port, etc.
- *Participatory management planning*: Planning of fisheries management policies and strategies at the national level, and detailed planning for management zones or areas with input from stakeholders (provinces, districts and fishers).
- *Establishing a regulatory framework:* The management plans need to be supported by appropriate legal instruments by means of which the plans are implemented. These legal instruments detail all the control mechanisms available for fisheries management including, but not limited to:
  - *Input controls:* Such as access (number of fishers, number of vessels by fishery), licences, closed seasons, gear restrictions, vessel limitations, area restrictions (Protected Areas), VMS requirements, and vessel identification.
  - *Operational and output controls*: Such as species and catch limits, by-catch limits, reporting requirements, air surveillance, sea patrols/inspections, boarding, logbooks, dockside monitoring, observers, port inspections, and catch documentation schemes.
- *Implementation*: This includes such measures as:
  - Participatory community-based management (CBM);
  - "Preventive" MCS activities to encourage voluntary compliance;
  - Public awareness and education campaigns;
  - Assistance to small scale fishers for supplemental livelihood development to reduce coastal area pressures;

  - Full enforcement to ensure compliance by those minority of fishers that persist in ignoring the law.

Similarly, MCS involves:

- Data collection and analyses for both operational planning and execution, as well as management planning, defined as monitoring (M);
- Involvement in the participatory management planning to include discussions on appropriate implementing mechanisms;
- Development and approval of appropriate and enforceable legislative instruments and control mechanisms such as licences, permitted fishing gear, seasons, vessel sizes, fish sizes, species, catch limits, by-catch limits, and area controls, or other restrictions to support the management plan, referred to as the control mechanisms (C); and
- Implementation of the plan through "preventive" and "deterrent" MCS techniques, included in the idea of surveillance (S).

Unfortunately, not all fisheries administrators understand MCS, or its critical role as an implementing mechanism for fisheries management. Some view arrests as the only relevant indication of the effectiveness of MCS efforts. The real indicator for MCS is the level of compliance, and this is governed by many factors, *e.g.* the number of fishers; the number of vessels; effort and area coverage of patrols; results of patrols, increase in voluntary compliance, etc.

Effective MCS involves a two-pronged, parallel approach. The *preventive approach* is to encourage "voluntary compliance" through understanding and support for the management strategies and this includes:

- Enhancement of community/fisher awareness and understanding of management practices and MCS through seminars, public awareness and information, education, and communication campaigns;
- Participatory management development to promote ownership of the management regime and input into the regulatory/control aspect of management (laws and regulations) in preparation for acceptance by the fishers of their joint "stewardship" role for the management of their fisheries in partnership with government;
- Peer pressure towards voluntary compliance and support for the management regime;
- The institution of accurate and verifiable data collection regimes; and

• Surveillance and verification for compliance.

The parallel approach of *deterrent/enforcement* MCS is necessary to ensure compliance by fishers who resist the regulatory regime to the detriment of both the fishery and the economic returns to their fellow fishers. Deterrent and enforcement include inspection, investigation, prevention and court proceedings to enforce the law. Voluntary compliance will fail if stakeholders see non-compliant fishers successfully evading the law and receiving economic returns from their illegal activity, at the expense of the fishers who comply with all requirements.

Namibia has used an effective mix of input and output controls to build a very successful fisheries management regime in the short period since independence was declared in 1989.

MCS Spatial Components

There are three main spatial dimensions or components to MCS: land, sea and air. The proper configuration of these spatial components for a given system will depend on such factors as cost, commitment, and organizational structure (national, subregional, or regional). All three components can now effectively use satellite technology (*e.g.* for vessel licensing, data collection and enforcement operations).

The *land* component of an MCS system serves as the base of operations, the co-ordinating centre for all MCS activities from which governments can regulate the deployment of resources to best address changing situations. The land component also entails port inspections, dockside monitoring, and the monitoring of transshipments and trade in fish products to ensure compliance with relevant rules.

Governments must also undertake a variety of land-based activities in order to carry out their responsibilities as flag States of vessels that may be fishing in remote areas, including on the high seas and in waters under the jurisdiction of other States. New technology has allowed States to link the land components of their MCS system to those of other States on a regional or subregional basis, which can greatly foster co-ordinated and responsible management.

The *sea* component includes MCS activities undertaken in marine areas under the jurisdiction of a State and may also cover high seas areas. Technology that comes into play can include radar, sonar and vessel platforms. While physical presence through at-sea patrols remains a fundamental part of this component (necessary for arresting violators and securing evidence), the costs involved have prompted a growing number of States to employ "no force" surveillance techniques. Examples of such techniques include the placement of independent observers on board fishing vessels, national or regional vessel registers and VMS requirements.

Developing coastal States are also requiring the flag States of vessels that wish to fish in their EEZs to ensure that those vessels do not violate the terms of the access granted. However, an at-sea presence remains necessary for the active enforcement of management measures to protect marine fisheries resources.

The *air* component covers the air and space equipment (aircraft, satellites, etc.) used in MCS activities. The flexibility, speed and deterrence of air and satellite based surveillance systems make these very popular tools for fisheries management. The air component provides for the rapid collection and dissemination of a wide range of information, including fishing vessel identification and reported fisheries data. Air, satellite or VMS surveillance can often provide initial information regarding fishing activity; they can also serve as first indicators of potential illegal activity and can thus trigger further MCS action. The cost of the air component is directly related to the sophistication of the technology used. Because governments can use air, satellite, video and advanced digital photographic technology, and VMS surveillance technology to address tasks beyond MCS (*e.g.* environmental and coastal zone monitoring, customs, immigration, and search and rescue), this equipment can improve the cost-effectiveness and efficiency of an integrated ocean management programme.

The United States of America has used all these tools in the past and currently has a multi-agency approach to MCS as noted.

The United States EEZ is reputedly the largest in the world at 3.36 million square miles (8.684 million km?). It hosts some 100 000 commercial fishing vessels. The 1999 fisheries resulted in landings of 9,339,034,000 pounds (4.2451 million metric tonnes) valued at US$ $3.4671 billion. Pollock, menhaden and salmon in that order accounted for 40% of the landings of the top ten species; while shrimp, crab and salmon brought in the highest revenues.

The management of fisheries is shared with the federal government and the states with the latter having management authority to 3 nm. The 200 mile EEZ was declared in the late 1970s at which time the USA permitted foreign fishing vessels in their waters under very strict conditions of entry and exit reporting, licensing, observers, and access only to specified windows for fishing with further gear, catch and effort controls. Foreign fishing ceased in USA's waters in the early 1990s, as the United States fishing fleet increasingly encountered pressures of over capitalization similar to other States. MCS control mechanisms for the domestic fisheries include the use of: licenses, quota, season, gear and area controls with appropriate reports and landing verifications, and observers for specific fisheries.

Vessel monitoring system technology is in use in the USA for fisheries vessel tracking on approximately 500 vessels in Hawaii, New England and Alaska, with plans for expansion for another possible 3,330 vessels in the Atlantic and Gulf of Mexico.

MCS Agency Responsibilities:

Each individual State is responsible for its 3 nm zone. The National Marine Fisheries Service (NMFS) under the National Oceanic and Atmospheric Administration (NOAA) conducts shore monitoring and at times places personnel aboard multi-agency assets for MCS, *e.g.* air patrols and sea patrols.

The United States Coast Guard (USCG) is the key agency responsible for enforcement within the EEZ and their air and sea assets are multi-tasked with fisheries being one of the secondary tasks. Coast Guard vessels spent 92 864 hours in fisheries related operations in 1999. USCG air hours for fisheries for 1999 were 10,527 hours. (This does not include the individual State efforts and dedicated efforts for MCS in their coastal waters.) The NOAA Office for Law Enforcement has five divisions based out of Silver Springs, Maryland and supports 50 field offices throughout the United States and Territories. It is the acknowledged leader and expert in the field of marine resource enforcement in the United States. The United States is developing an integrated national system so all fisheries that may be subjected to VMS requirements will be centralized for analysis, planning, and operations.

Land-use Issues in Aquaculture

Regions suitable for land-based aquaculture are limited by the right combination of climate, soil type, geomorphology, water availability, and access to other resources such as labor, markets, feeds, and larvae or juveniles for stocking. Nevertheless, land use per se is seldom a limiting factor in aquaculture development. This is particularly true in the United States, where relatively few locations are significantly developed into aquaculture.

The Yazoo-Mississippi River alluvial valley in northwest Mississippi is the most intensely developed aquaculture region in the United States, with aquaculture ponds occupying more than 45,000 ha of land. Yet ponds account for only about 2% of land use in the region. Even in sub-watersheds within the Yazoo basin with more intense development, ponds account for only 10 to 20% of the watershed.

Aquaculture development in the United States is more likely constrained by water availability, regulatory constraints on land development, or the value of land for alternative uses than by the availability of suitable land for facility construction.

At times, availability of the right combination of resources leads to localized overdevelopment of aquaculture. In this respect the problem is not necessarily the physical concentration of facilities, but rather ecological imbalances associated with the inability of the local ecosystem to provide support services, such as adequate water supply or waste treatment.

A good example is the study of the ecosystem support area needed for semi-intensive shrimp farming in the Bay of Barbacoas area of Colombia, South America. The estimated local ecosystem support area needed to assimilate waste nutrients from shrimp farms closely matched the mangrove area available to provide that service. Local capacity for waste treatment was used to its fullest extent and additional development could lead to self-pollution of the water supply, spread of disease, and other environmental impacts that have been associated with the collapse of shrimp farming in other overdeveloped areas.

Most problems associated with land use in aquaculture are related to the value of alternative uses of the land. These conflicts arise when aquaculture must compete with other land uses. Land-use conflicts tend to be especially acute in coastal regions, which are limited in area, have proportionately large areas of sensitive and environmentally valuable ecosystems, and are intensively used for a range of often conflicting purposes by dense human populations. Coastal regions also tend to be more commercially developed and the land more costly, especially in the United States.

Clearing mangrove forests to build ponds for shrimp and fish culture has been one of the most contentious land use conflicts in aquaculture. Mangrove ecosystems have high inherent aesthetic and environmental value. They protect coastlines from erosion and flooding and provide habitat and food for a tremendous variety of aquatic, terrestrial, and avian species.

Mangroves also act as natural purification systems for waters interacting at the interface of freshwater and marine environments. They are also convenient—although far from ideal—locations for brackishwater aquaculture ponds, especially for growing shrimp. In the 1980s and early 1990s, when global shrimp aquaculture was expanding rapidly, thousands of square kilometers of ponds were built in mangrove areas. Much of this land had been cleared previously for other uses, but the destruction was often attributed to shrimp farming. Although environmental advocacy groups overstated the involvement of shrimp farming in the historic loss of mangrove, worldwide awareness of the problem resulted in stricter governmental oversight of the use of mangrove forests and stimulated significant changes in shrimp production practices to protect these valuable ecosystems.

Aquaculture can affect sensitive ecosystems other than mangroves. Marine benthic communities may be impacted by net-pen aquaculture, and construction and operation of freshwater ponds can destroy sensitive wetlands. Certain types of open-water molluscan shellfish culture—usually viewed as an environmentally benign endeavor—can damage seagrass beds due to siltation and the mechanical effects of certain culture practices.

Marine aquaculture also faces unique conflicts unrelated to the physical area occupied by the facility or the impacts of culture activities. Conflicts may arise from competition with other users of the area or region, including activities such as tourism, shipping and navigation, capture fisheries, and urban development. Conflicts may even arise from esthetic considerations, including the effect of aquaculture development on the scenic quality of an area (the *viewscape*) and disturbances related to increased activity and noise associated with culture practices.

Nearly all problems arising from land-use conflicts in aquaculture can be avoided by prudent site selection. Better management practices for facility site selection.

Agriculture and Water Resources

Cultivated systems, including freshwater aquaculture, now cover 24% of the terrestrial surface of the planet. Most of the increase in food production in the last 50 years has been derived from agricultural intensification, not an increase in cropland area. Intensification has increased pressure on water resources (both for supply and waste treatment) and fisheries and cropland resources for feed, and it requires increased energy resources, nearly all derived from fossil fuels. Freshwater is a scarce resource on the planet, with less than 1% of total freshwater resources available for human use.

Renewable freshwater resources available for human use are derived from precipitation that falls on land. This precipitation runs off (including infiltration to groundwater) or is lost by evapotranspiration from land and water surfaces and through plants. Humans use about 10% of renewable freshwater supplies, although some countries use more than 100% of renewable supplies.

About 40 to 50% of freshwater runoff is appropriated for human use. The available supplies of global freshwater are used by agriculture (75%), by industry (20%), and for domestic purposes (5%).

Aquatic resources are impacted by numerous human activities. Running waters are exploited to supply irrigation and drinking water, generate electricity, and receive wastes. Running waters are also affected by land use in adjacent watersheds.

Threats to running waters include alteration of habitat; changes in water chemistry; and changes in biodiversity caused by land transformations, channelization, and industrial or urban development. Standing waters provide water for agricultural, industrial, and domestic uses. Threats to standing water ecosystems include eutrophication, chemical contamination, and the introduction of invasive species.

Water use in Aquaculture

Water is an obvious and essential resource for aquaculture. No single factor influences the success of aquaculture more than availability of an adequate supply of good quality water. Availability of freshwater increasingly constrains all forms of human development, so aquaculture must compete with other uses, including irrigated agriculture, industry, and urbanization.

Competition for finite water supplies is particularly acute in arid regions or where freshwater is supplied primarily from groundwater resources. Sparing water for other uses by adopting conservative water use in aquaculture is therefore a social obligation. Water use is also inextricably tied to issues of waste discharge because water input and discharge are variables on two sides of the hydrological equation: increasing water input volume will increase water volume discharged.

Water use in aquaculture may be classified as either total use or consumptive use. Total water use is the sum of all inflows (precipitation, runoff, seepage inflow, and management additions) to production facilities. Depending on the type of culture system, a portion of the water entering the facility passes downstream in overflow or intentional discharge and is available for other purposes (e.g., ecosystem support, irrigation). For example, essentially all water flowing into flow-through systems is discharged, whereas some ponds have no discharge at all. Consumptive use can be defined as water used in aquaculture that is not available for other purposes. Consumptive water use includes water lost by evaporation and seepage from an aquaculture facility, withdrawal of fresh groundwater, and water removed in biomass of aquatic animals at harvest. Water in harvest biomass averages about 0.75 L/kg, a minor quantity compared to other uses.

Potential Environmental Impacts of Aquaculture

The environmental effects of aquaculture are often described in terms of the sources (type and quantity) or causes of a potential effect, and this convention will be followed here. Impacts are potential because effects depend on characteristics of the environment in which aquaculture production systems are embedded. For example, releasing waste into a closed basin will have a qualitatively greater (more adverse) impact that releasing the same quantity and type of waste into an open basin.

Contrasting to broad issues related to resource use, many of the problems commonly attributed to aquaculture—such as the adverse effects of habitat conversion, waste discharge, consumptive water use, and release of chemicals and antibiotics—are related to facility operations and can be addressed on shorter time scales through technology-based approaches. Solutions to problems of this nature are the focus of this book.

Habitat Conversion

Ponds are the most common aquaculture production system in the United States and the world. In most cases, pond construction requires conversion of terrestrial or wetland habitat to aquatic habitat. Aquaculture ponds must be located near sources of water, so they are commonly constructed in coastal or riverine floodplains. Thus, pond construction may convert floodplain land or wetlands to ponds.

Pond construction in or near coastal wetlands, particularly the conversion of tropical mangroves to shrimp ponds, is controversial. Coastal wetlands are multiple-use, mostly open-access resources that are increasingly threatened by a wide range of development pressures, including aquaculture. Coastal wetlands provide a number of ecosystem services, including waste treatment, storm protection, food production (especially as a nursery for many aquatic species), recreation, and a source of raw materials.

The environmental impact of pond construction in coastal wetlands can be described in terms of the degree to which these services are lost by conversion to ponds. The total economic value (including the value of nonmarketed ecosystem services) of sustainably managed mangrove wetlands often exceeds the value associated with conversion to shrimp ponds. Coastal development, including aquaculture, in tropical areas can adversely affect adjacent, light-sensitive ecosystems, such as coral reefs and seagrass meadows.

Approximately 35% of mangroves have been lost in the last 2 decades. Estimates of the relative contribution of aquaculture to the loss of coastal mangroves are uncertain, although less than 10% of the global loss of mangrove area has been attributed to conversion into shrimp ponds. Nonetheless, conversion to aquaculture ponds can be the major component of mangrove loss in some coastal areas, particularly in southeast Asia. Mangrove wetlands are increasingly seen as having intrinsic value, being poor sites for pond construction, and undesirable sources of social conflict when aquaculture projects are developed. Pond aquaculture in coastal wetlands is much less common than in the past. In many places in Asia, mangrove clearing has been banned and rehabilitation and restoration programmes are active.

Waste Loading

Nearly all aquaculture production systems produce waste nutrients and organic matter. Depending on the production method, organic fertilizers (manures and agricultural byproducts), inorganic fertilizers, and nutrient-dense feeds are applied to production units in most aquaculture facilities to promote growth of finfish, crustaceans, and other species. Approximately 70 to 80% of nutrients in feed are released as waste into the culture unit in aquaculture production systems where cultured animals are fed.

The degree to which the culture unit is open to the environment will determine the proportion of waste excreted by cultured animals that is released to the environment. For example, ponds and recirculating systems are operated with a long hydraulic residence time, so a very small proportion of nutrients are released to the environment. In contrast, culture systems with a more direct hydrological connection with the outside environment, such as flow-through systems and net pens, will release a large proportion of waste nutrients and organic matter.

The ecological effects of discharged nutrients will vary depending on production intensity, facility density, the hydraulic retention time and trophic state of receiving waters, and dilution rate. In relatively closed production systems such as ponds, feeding invariably results in the eutrophication of pond water. In more open systems, the effects of nutrient discharge are far less predictable. If the flushing time of the water body is less than the generation time of phytoplankton, an increase in limiting nutrient concentration can result in eutrophication. It is important to emphasize the point that phytoplankton respond to changes in nutrient concentration, particularly of those nutrients that tend to limit production (phosphorus in freshwater and nitrogen in seawater), not to increases in nutrient mass loading per se.

Potential adverse environmental effects attributed to organic waste loading from aqua-culture include increases in phytoplankton density in response to elevated nutrient concentration, increased variation in dissolved oxygen concentration in response to increased phytoplankton density, increased frequency of dissolved oxygen depletion, reduced dissolved oxygen concentration caused by the discharge of effluent with high biochemical oxygen demand, and changes to benthic communities caused by localized sedimentation of suspended solids. Nutrients discharged from aquaculture facilities have not been linked to the development of harmful algal blooms.

In many coastal areas, discharges from aquaculture production facilities represent one of many nutrient sources contributing to eutrophication. Truly addressing problems related to eutrophication therefore requires that the relative contribution of aquaculture be evaluated in terms of all sources and considered in the context of a watershed or whole-basin approach. In certain locations, particularly closed basins, aquaculture operations can be the primary contributor to localized eutrophication. The effects of effluent discharge tend to be localized near the discharge point. The susceptibility of receiving waters to eutrophication depends on trophic status prior to enrichment. For a given level of nutrient loading, oligotrophic waters are more susceptible to trophic state changes than mesotrophic waters.

Most discussions of waste loading from aquaculture consider only the components of the waste stream, which includes nutrients, organic matter, potential pathogens, chemicals or therapeutants, and escaped fish. However, they do not consider how the environment responds to waste loading. Often missing is the notion of environmental carrying capacity, which requires an understanding of the waste assimilation capacity of the environment and consensus on an acceptable level of environmental change.

Aquatic environments have an inherent capacity to assimilate nutrients through well-known biological and physicochemical processes. In some areas, aquaculture facilities are sufficiently concentrated and production intensity is sufficiently great that the majority of the assimilative capacity of receiving waters is used by aquaculture. Appropriation of the waste treatment ecosystem service by aquaculture has been criticized on the basis that the cost of waste treatment is externalized by aquaculture operations and born by society. In general, nutrient enrichment is perceived to have negative consequences, although in some environments, eutrophication may be beneficial and perceived as desirable. Changes in trophic state have no inherent value; the acceptability of the change is a matter of societal values and policy.

Benthic Impacts

Organic wastes produced in open aquaculture production systems can accumulate on the sediment in the vicinity of culture units. Soil eroded from pond embankments and organic matter in pond water can settle in receiving water bodies adjacent to discharge points. Organic matter discharged from flow-through systems can accumulate near the discharge point, leading to localized suppression of dissolved oxygen concentration and changes in benthic communities. Depending on current speed and water depth, organic matter mainly derived from feces and uneaten feed can accumulate on the sea floor beneath net pens. Sedimentation rates within net-pen facilities are reported to range from about 15 to 100 g of total volatile solids (TVS)/m^2 per day, with most estimates between 25 and 50 g TVS/m^2 per day. The accumulation of organic matter can extend from 145 to 205 m downcurrent from the perimeter of net-pen facilities, although significant effects are usually restricted to less than 60 m from the perimeter. Of all the potential environmental impacts of salmonid net-pen aquaculture, changes to the sediment beneath net pens are considered to represent the greatest risk to the environment, although these changes are temporary and largely reversible.

If the rate of organic matter sedimentation exceeds the rate of decomposition, organic matter will accumulate. Organic matter accumulation will depend on loading rate, temperature, oxygen supply, current velocity, and other physical and chemical factors. The deposition of organic matter creates a dissolved oxygen demand for decomposition that may exceed supply. In this case, the redox potential will shift to lower (more negative) values, indicating anaerobic conditions. Mats of the sulfide-oxidizing filamentous bacteria *Beggiatoa* may cover the sediment surface. Reduced substances such as ammonia, dissolved phosphorus, methane, carbon dioxide, and especially in marine waters, hydrogen sulfide will diffuse into the overlying water. Some of these substances are potentially toxic to benthic invertebrates and cultured fish, but rarely accumulate to levels that will affect fish culture performance.

Anaerobic conditions in the sediment will also cause a shift in the assemblage of benthic invertebrates towards a less diverse community dominated by pollution-tolerant species. The effects described here tend to be localized around effluent discharge points and within 25 m of the perimeter of net-pen farms (Karakassis *et al.* 2000; Pearson and Black 2001). These effects are also more pronounced for poorly flushed sites (current velocity <10 cm/ second) with fine-grained sediments. In well-flushed sites (>50 cm/second), the abundance and diversity of benthic infauna frequently increases. Chemical remediation of marine sediments beneath net pens occurs within months to 1 year, but biological remediation may require 2 to 3 years in some sites, largely depending on water temperature, current velocity, and concentration of sediment organic matter. In general, the effects of sedimentation are localized and reversible.

Sedimentation can also occur as a result of alteration in hydrodynamics caused by the deployment of aquaculture structures such as rafts, longlines, and cages. These can reduce current velocity in the lee of culture facilities, creating a depositional environment. In addition, feces and psuedofeces produced by shellfish can accumulate in the vicinity of shellfish rafts. Hydrodynamics can be altered by material that accumulates near effluent outfalls, which can reduce stream depth and obstruct water flows, increasing the potential for flooding. Some metals will accumulate in sediment, but there is no evidence that these accumulations have had negative impacts on benthic invertebrates or rates of biogeo-chemical transformations. Zinc is added to fish feeds as an essential nutrient, and elevated concentrations have been measured in sediment beneath net pens. Copper is used in antifouling paints and net treatments and can accumulate in sediment beneath net pens. In marine environments, the accumulation and bioavailability of copper and zinc depend on sulphide concentration, with reduced bioavailability with elevated sulphide concentration. Concentrations decline during site fallowing.

Chemical Pollution

As with other forms of agriculture, chemicals are used in aquaculture for a broad diversity of purposes. Perhaps most importantly, pesticides, disinfectants, and antibiotics are used for disease treatment and pest management. Chemicals are also used for soil and water treatment, enhancing natural productivity, feed manufacture, control of reproduction, growth enhancement, transportation, and processing. Major classes of chemicals include therapeutants (including antibacterial agents), disinfectants, anesthetics, hormones, feed additives, antifouling paints, herbicides, and pesticides. Soil and water are treated with chemicals such as inorganic and organic fertilizers, liming agents, flocculating agents, and bacterial amendments. Chemical and drug use are highly regulated in the United States, and even usage of legally registered substances is relatively uncommon for reasons of economics. This is not necessarily true elsewhere, however. On average, a shrimp farmer in Thailand uses 13 chemicals, with four pesticides and disinfectants, and three soil and water treatment products. Thai shrimp farmers use an average of one antibiotic per farm, most commonly fluoroquinolones.

Chemicals used in aquaculture can be divided into three groups based on the different ways they affect the environment. One group of chemicals encompasses those with acute or chronic toxicity to nontarget organisms, including cultured animals. Another group are the antibiotics that are released to the environment. These can lead to the development of resistant strains of pathogenic bacteria and alter microbial community composition, affecting biogeochemical processes. The third group are nutrients that can lead to eutrophication if discharged, a particular concern for light-sensitive seagrass and coral reef ecosystems often found adjacent to coastal shrimp ponds. Chemicals can also be grouped into three categories based on the environmental or human health risk associated with their use. The first group includes chemicals with a high risk associated with their use, such as chloramphenicol, malachite green, and organophosphates. Chemicals in a second group can be used safely if applied according to label directions, but may present an environmental or human health hazard if used incorrectly. A third group of chemicals can be used safely at most locations, but may present an elevated risk at certain sites because of particular characteristics of that site.

Chemicals have various fates when released from aquaculture production systems to the environment, so environmental persistence is variable. The persistence of chemicals in the environment depends on physicochemical factors that affect solubility and reactivity (especially temperature, pH, dissolved oxygen concentration, and light intensity) and biological factors, especially the type and density of microorganisms. Some chemicals can accumulate in sediments. Other chemicals dissipate or degrade by physicochemical reactions such as photooxidation or adsorption to sediment minerals or organic matter. Others are transformed or degraded biologically by reactions mediated by microorganisms. The environmental toxicity of reaction products can be greater or less than that of reaction substrates. In general, but with some important exceptions, most chemicals used in aqua-culture do not persist in the environment.

The environmental and human health issues associated with chemical use in aquaculture were summarized by GESAMP. These include persistence in aquatic environments, residues in noncultured organisms, toxicity to nontarget organisms, stimulation of bacterial resistance, effects on sediment biogeochemistry, nutrient enrichment, health of workers exposed to chemicals, and residues in seafood that affect product safety. Many of these effects can disturb natural aquatic communities, altering community structure and reducing biodiversity.

There is a strong disincentive to use many chemicals in aquaculture because use increases production costs and some chemicals may be toxic to cultured aquatic animals. Chemical use also presents a food safety problem by increasing the risk of consumption of aquatic animals containing chemical residues. Improper use of antimicrobials can stimulate the development of drug-resistant forms of microbial pathogens in cultured animals, water, or sediment. Chemicals can also bioaccumulate in nontarget organisms and become biomagnified as they move through aquatic food webs to higher trophic levels.

In the particular case of salmonid net-pen aquaculture, there has been concern about the use and fate of pesticides and drugs applied to control infestations of various species of parasitic copepods known collectively as sea lice and the antibiotics used to control bacterial diseases. In some countries, sea lice are controlled with bath treatments of nonspecific and broad-spectrum organophosphorus and pyrethroid pesticides and with the antibiotic ivermectin incorporated into feed. The chemicals used in bath treatments disperse rapidly and are usually not detected beyond 25 m from a net pen, although these chemicals have the potential to affect nontarget organisms, particularly crustaceans. Research on the fate of antibiotics used to control bacterial diseases indicates that antibiotic residues will accumulate and persist in sediment beneath cage sites and have been measured in nontarget organisms. There is concern about the increased risk of the development of drug-resistant strains of bacteria that are pathogenic to fish, particularly when antibiotics are used prophylactically. Similar concerns have been expressed about the development of drug-resistant strains of bacteria in shrimp farming.

Salinization

Salinization refers to increases in the salinity of soil, surface water, and groundwater. Excessive salinization of water and soil can limit the use of surface waters for irrigation, reduce crop yields, displace salt-intolerant crops, and limit the use of water for domestic purposes. Salinization is primarily a concern with coastal pond aquaculture, where water is pumped to maintain appropriate salinity in ponds for penaeid shrimp culture. However, salinization has become a concern with the increase in inland aquaculture of marine shrimp. These facilities are supplied with brine transported from coastal waters or saline groundwater from wells, and salinization can occur when this water is discharged to inland streams. Salinization can also occur when accumulated sediment is removed from coastal aquaculture ponds and disposed in freshwater areas. Salt contained in the sediment leaches out following rainfall and can cause salinization.

The discharge of water from coastal or inland brackishwater ponds into freshwater bodies can impair surface water quality to limit their use, especially for agriculture. In sites with improperly compacted soils, water can seep from these ponds into underlying freshwater aquifers. Alternatively, excessive pumping of groundwater can result in saltwater intrusion of aquifers, thereby limiting use. Excessive pumping of groundwater to control salinity in coastal aquaculture ponds has led to salinization of groundwater aquifers and land subsidence and consequent damage to infrastructure in Taiwan. Pumping groundwater was widely used to manage salinity in brackishwater ponds in the past, but the practice is now banned or restricted in many places.

Pathogen Transmission

In aquaculture production systems, animals are usually held at much greater densities than in the natural environment, a condition that facilitates the transmission of pathogens among cultured animals. Elevated production intensity can lead to environmental conditions in the production unit that stress cultured animals, potentially increasing the susceptibility to infection by pathogens that may lead to disease. Thus, the conditions that promote disease are more prevalent in aquaculture production systems than in the natural environment. Intensive culture conditions can also cause amplification of pathogens that are ubiquitous in the environment.

Pathogens are a constituent of water that is discharged from or flows through aquacul-ture facilities. Water exchange in open production systems facilitates the transmission of pathogens to the environment, specifically between cultured and wild fish. There are numerous examples of diseases caused by the discharge of pathogens from aquaculture production facilities.

Wild fish also serve as a reservoir of pathogens for the infection of cultured fish. The transfer of pathogens to wild fish can potentially affect the susceptibility of wild fish to disease and, in extreme cases, could affect wild fish abundance and local biodiversity.

The discharge of pathogens does not necessarily lead to disease. Infection with a pathogen is a much more common occurrence than disease. Infectious diseases result from the interaction of the health and immunological status of the host, the dose and virulence of the pathogen, and environmental conditions that affect host-pathogen interaction. The health of wild fish populations is influenced by natural factors, such as genetics, nutrition, environmental conditions, and the biology of pathogen and host organisms. Fish health is also affected by anthropogenic factors that put pressure on wild fish populations, including climate change, floods, drought, impoundments, dams, chemical contaminants, and fishing pressure.

The risk of disease caused by a particular pathogen is affected by the characteristics, distribution, survival, and fate of pathogens in the environment; factors affecting the route of pathogen transmission; the characteristics of pathogen exposure; the effect of environmental stressors; and the susceptibility of wild hosts to infection that can lead to disease. In general, the relative importance of each factor is difficult to understand and assess in wild populations, leading to a high degree of uncertainty associated with the risk of transfer of pathogens between cultured and wild fish populations.

Wild fish can be affected by the introduction of new pathogens attendant with the introduction of native or nonnative species to an area for culture. These pathogens can infect wild individuals of the same or different species. Pathogens can also be introduced to wild fish through transportation of native or nonnative species that are not intended for aquaculture.

Interactions of Escaped Fish with Wild Populations and Natural Ecosystems

Aquaculture production facilities are embedded in natural ecosystems with varying degrees of intimacy, depending on characteristics of the culture system. The hydrological connection to natural ecosystems can be remote and intermittent (e.g., recirculating aquaculture systems) or intimate and continuous (e.g., shellfish and net-pen systems). The risks of containment failure also vary with culture system type. As such, the probability of occurrence of escape is variable. Escapes can be defined as the unintentional and unplanned release of cultured animals or their gametes or offspring to the environment. Escapes do not include intentional releases of cultured animals for stock enhancement or ranching.

Mechanisms that lead to the escape of cultured organisms can be grouped into four classes. In each case, the mechanism leads to a breach in the barrier between a culture unit and the environment, allowing escapes to occur. The causes of containment failures that can lead to escape include 1) catastrophic natural events, 2) human error associated with facility operation, 3) vandalism or poaching, and 4) attacks by nuisance or predatory wildlife. Culturists have no control over natural disasters, but facilities can be constructed to withstand or minimize the risk of escapes by other mechanisms.

Escape of cultured animals is not desirable from an economic standpoint because escapes represent a loss of potential revenue. Therefore, there is a strong incentive to construct and manage production units to minimize losses from escapes, disease, poaching, and other causes. Above and beyond the undesirable economic impact on producers, escaped culture organisms can alter the physical environment and the structure of ecological communities, cause genetic changes in stocks of wild conspecifics, introduce parasites and diseases, and cause various socioeconomic impacts.

Escaped individuals may interbreed with natural populations, although interbreeding requires the satisfaction of several criteria. First, cultured organisms must be sexually mature or, if not mature, they must survive to maturity. The timing of the escape event must also coincide with the natural breeding season of the wild population; in other words, sexually mature individuals in the wild population must be receptive to escaped fish.

Intraspecific hybridization between escaped and wild conspecifics can reduce the genetic variance, and presumably fitness, of wild populations through outbreeding depression. This may increase the vulnerability of natural populations to environmental change from the loss of genetic differences between cultured and wild (natural) populations. The effect of inbreeding will depend on the degree of local adaptation. Geographically structured populations are more susceptible to the effects of interbreeding than undifferentiated stocks.

The principles of interaction between wild and escaped fish have been reviewed by Youngson *et al.* (2001). The overall concern is related to a reduction in fitness and a reduction in the effective population size of wild populations.

First, the genetic profile (*i.e.*, allele frequencies) of wild populations can change from the transfer of genes from cultured stocks into wild populations. Second, the genetic profile of cultured stocks may contain alleles that are rare or unusual in wild populations but were selected on the basis of performance traits, and that are transferred to wild stocks. Third, the abundance of wild populations may decrease if escaped fish displace wild fish, but then demonstrate low reproductive fitness. Although all of these mechanisms of interaction are plausible, demonstrating a measurable effect on the fitness of wild populations is difficult.

The consensus view is that the risk of adverse genetic effects on natural populations of conspecifics is not as severe as potential ecological effects of escapes.

The intensity of the ecological impacts of escapes is a function of the number of organisms that escape, the timing of an escape event, the biology and life history of the escaping species, interactions of escaped organisms with wild populations, and ecological characteristics of the host ecosystem. Even nonreproductive animals can have ecological impacts if they escape in sufficient numbers.

Potential adverse impacts of escapes on natural populations may include any or all of the following :

1. Competition with natural populations for habitat, food resources, nesting or spawning sites, or mates;
2. Predation on endemic fish populations if escaped fish are piscivores;
3. Predation on endemic fish populations if escaped fish attract predators;
4. Transmission of pathogens or parasites with deleterious effects on wild populations;
5. Amplification of the effect of endemic pathogens;
6. Alteration of habitat, such as changes in turbidity and substrate; and

7. Colonization of habitat, with negative effects on wild populations, and in the worst case, displacement of native species.

Overall, the potential outcomes of escapes may include a reduction in fitness in wild con-specific populations and a reduction in fitness in other wild populations.

Many of the adverse impacts described above are exacerbated when invasive species escape from aquaculture facilities. Although only one of many pathways for the introduction of potentially invasive species, aquaculture has played an important role in the introduction of nonnative species, especially in freshwater. The majority of introductions of nonnative species since the 1970s can be attributed to aquaculture and arguably the introduction and establishment of invasive species from aquacul-ture has had the greatest adverse effect on the environment of all potential impacts. In any case, the effects of escapes by cultured species on biodiversity are difficult to partition from the effects of habitat degradation or loss and other causes of biodiversity reduction.

Collection of Wild Aquatic Animals for Spawning or Stocking

There is a long history in aquaculture of capturing wild larvae and juveniles for stocking into culture units for growth to market size. The culture of milkfish, mullet, penaeid shrimp, and many other species was long dependent on wild stocking material. Capturing wild juveniles is usually an interim step in the development of a species for culture. Once the techniques for spawning and early larval rearing are developed to close the life cycle, juveniles can be reliably produced in hatcheries, allowing the development of a sustainable aquaculture industry for that species. Most species in freshwater aquaculture are now spawned in captivity. The dependence on wild-caught stocking materials is becoming much less common, although reliance is still strong in brackishwater and marine aquacul-ture. The capture-based aquaculture for eels, groupers, tunas, and yellowtails remains dependent on capture of wild juveniles.

There are several problems related to the use of wild juveniles for stocking. Collection of juveniles may place additional pressures on wild stocks of the target species, reducing recruitment to natural fisheries if density-dependent mechanisms regulate abundance. Furthermore, depending on the homogeneity of aggregations of juveniles, individuals of nontarget species of commercial or ecological importance may be collected incidentally as by-catch, placing pressure on wild stocks of those species. For example, in the Philippines, milkfish fry constitute only 15% of collections and the 85% of by-catch is discarded. The effect of capture of wild fish as juveniles on target or nontarget wild populations will depend on the size- and age-structure of the population, population size, age at maturity, natural and fishing mortality rates, and quality and availability of suitable habitat. Selective collection of wild juveniles can also have a negative effect on fish community structure, depending on the trophic role of the species and the complexity of the community structure. Depending on the gear employed, juvenile collection can result in physical disturbance of the substrate, impacts on benthic communities, and in the worst case, destruction of habitat. Finally, wild seed are undesirable from a biosecurity standpoint because the health status of wild-caught juveniles is questionable. Shrimp producers now perceive that postlarvae produced in a hatchery are less susceptible to important diseases than wild postlarvae.

In the culture of some species, sexually mature broodstock are collected from the wild and induced to spawn in hatcheries. Broodstock collection can also place pressure on wild fish resources. For example, in Thailand, Vietnam, and some other Asian countries, the availability of wild-caught black tiger shrimp (*Penaeus monodon*) for broodstock has declined drastically in recent years. Adult *P. monodon* broodstock are now rare in trawl catches. Capture of mature adults for use in shrimp hatcheries likely has been one of the reasons for the decline. This decline in *P. monodon* broodstock has been a major reason why shrimp producers in Thailand and other Asian nations have imported non-native, farm-reared broodstock of the Pacific white shrimp (*Litopenaeus vannamei*). This species has replaced much of the *P. monodon* culture in Thailand.

Predator Control and Other Effects on Wildlife

The concentration of potential food resources in aquaculture facilities provides attractive foraging opportunities for birds and other predators of cultured animals, which are viewed as a nuisance by aquaculture producers. The construction of aquaculture ponds has created new aquatic habitat while suitable habitat for wildlife with an aquatic orientation, especially migratory waterfowl, has decreased elsewhere. Ponds are attractive to waterfowl and migratory birds as sources of food and shelter. Aquaculture structures such as net pens and shellfish rafts create a complex, three-dimensional environment in the water column that attracts and aggregates wild fish, their predators, birds, turtles, and marine mammals.

Aquaculture producers use a range of nonlethal and lethal techniques to control preda-tion of cultured animals. Most methods are intended to harass wildlife and induce them to move elsewhere. Although lethal controls are used occasionally, it is unlikely that lethal control will have negative effects on wildlife populations, especially of those for which aquaculture has allowed an expansion of population abundance. Most wildlife is protected by international conventions and national regulations. It is possible that location in migratory pathways and near breeding grounds could cause disruption to wildlife populations, but it is unlikely that these sites would be selected or permitted for aquaculture.

There is some risk that wildlife may become entangled or trapped in nets, ropes, and floating or submerged structures for fish and shellfish. This could lead to drowning or starvation, but properly designed, constructed, and managed facilities do not present a major risk of entrapment or entanglement. In sum, there are few negative interactions with wildlife and the probability of negative consequences is very low.

Global Environmental Threats to Fisheries and Aquaculture

We live on a human-dominated planet, and the momentum of human population growth, together with the imperative for further economic development in most of the world, ensures that our dominance will increase.

Sustainable development and management of aquaculture and fisheries systems

The sustainable development and management of aquaculture and fisheries systems can only occur if these activities are well planned and integrated into the natural and social resource, ecosystems, and farming systems contexts of the larger global context of which they are a part. Population and natural resource constraints in a crowded future demand that aquaculture "fits" as a part of a larger strategy for the non-consumptive, multiple uses of water; and that fisheries be managed sustainably as part of the larger trends affecting the marine environment.

In the past 50 years there has been a massive migration—called the "greatest human migration of all time"—from rural, inland areas to the world's coasts, resulting in about 60 per cent of the Earth's people living within 100 km of the coast. Massive growth of coastal cities has raised issues of the future survival of coastal and estuarine ecosystems and habitats, and has put at risk the livelihoods of millions of people in traditional coastal communities that depend upon the sustainable capture and culture of aquatic living resources. Upwards of 35 per cent of the primary production of the temperate continental shelves is being harvested in fisheries, or discarded back to the sea as waste ("bycatch"). Coastal margins and oceans are among the most heavily used and modified areas of the planet, suffering disproportionate amounts of habitat destruction and pollution.

Humans have removed about 50 per cent of the world's mangroves. Human activities in the coastal zone deliver sewage, solid wastes, refuse (marine debris), sediments, dust, pesticides, and oil hydrocarbons to rivers, estuaries, and coastal areas. Assessing the range, magnitude and delivery of land-based sources of pollution to coastal oceans is a major global effort. It is estimated that about 80 per cent of marine pollution originates from land-based sources and activities. Inputs of nutrients to the coastal zone from development and agriculture have caused an increase in toxic algae blooms, some of which have caused human disease and neurological impairment. In Shanghai, China, an estimated 13,000 factories discharged 10.55 million tons of waste in 1989, nearly twice the amount in 1980; about one-fifth was toxic waste.

The Songhua River in China contains 10 tons of mercury and has waterborne mercury concentrations higher than those reported after the disaster in Minamata Bay, Japan. About one-third of China's coastal waters are polluted with oil; mercury and cadmium have been detected in 60 per cent and 33 per cent of seawater samples, respectively. In Poland, 92 per cent of the nation's rivers are "beyond classification", meaning that the level of pollution is greater than that described by any existing pollution category.

Humanity is the now one of the major driving forces in the Earth's hydrological cycle, using more than half of the world's freshwater run-off. Agricultural water use accounts for about 75 per cent of total global consumption, mainly through crop irrigation, while industrial use accounts for about 20 per cent, and the remaining 5 per cent is used for domestic purposes. Most of the world's rivers are dammed—there are 36,000 dams—and the number of dams is increasing.

Many major rivers (Colorado, Nile and Ganges) are so heavily used that little or no water reaches their deltas and the sea. Only 2 per cent of rivers in the USA run free. Major inland water areas (Aral Sea, Lake Chad, etc.) have nearly dried up or been greatly reduced to small lakes. It is estimated that two out of every three people will live in water-stressed areas by the year 2025. In Africa, it is estimated that 25 countries will be experiencing water stress (below 1,700 m^3 per capita per year) by 2025. About 450 million people in 29 countries suffer from water shortages. Clean water supplies and sanitation are major problems in many parts of the world, with 20 per cent of the global population lacking access to safe drinking water.

Water-borne diseases from fecal pollution of surface waters are a major cause of illness in developing countries. Polluted water is estimated to affect the health of 1.2 billion people, and contributes to the death of 15 million children annually.

Losses of biodiversity are occurring at alarming rates. Rates of loss are estimated at 100–1000 times greater than natural rates of extinctions. Extinction rates are even higher on remote islands, such as Hawai'i, USA. Due to the ease and rapidity of global transportation, the world's ecosystems are becoming "homogenized" due to "biological invasions": the exotic introductions of plants and animals, bacteria and other pests to natal ecosystems. After land transformations, exotic species impacts are the second major cause of species extinctions and population losses. About 11 per cent of birds, 18 per cent of mammals, 5 per cent of fish and 8 per cent of the Earth's plants are threatened with extinction, and exotic species have contributed greatly to their demise.

Given these immense global challenges, Vitousek *et al.* recommend:

- A slowing the rate of population growth since humanity's growth drives all resource use and waste generation,
- A reduction in the rate of human impacts since ecosystems can react to lower rates of impacts and stabilize with moderate levels of human-induced changes,
- Accelerating ecosystem level research and management understanding. Human impacts must be included in all analyses of global and ecosystem change. Ecological science must include development of new methodologies to assist humanity in this global crisis, requiring development of true interdisciplinary environmental scholarship, and education of the "informed generalist."

The Global Importance of Fisheries and Aquaculture

Fisheries play an important role in the world food economy. Fisheries are a source of employment for about 200 million people who depend directly upon ocean fishing for their livelihoods. Fish is the primary source of protein for some 950 million people worldwide and represents an important part of the diet of many more.

In less than 50 years, the world's average per capita consumption of fish has almost doubled. Globally, fish provide about 16 per cent of the animal protein consumed by humans, and are a valuable source of minerals and essential fatty acids. Fish is the primary source of omega-3 fatty acids in the human diet. Omega-3 fatty acids are critical nutrients for normal brain and eye development of infants, and have preventative roles in a number of human illnesses, such as cardiovascular disease, lupus, depression and other mental illnesses.

The reported production of fish for direct human consumption doubled between 1950 and 1970, and has stabilized since then at an average of 9.0 to 10 kg of fish per capita, notwithstanding world population growth. Fish consumption per person is expected to continue to rise. Supply will probably be limited by environmental factors, and a likely range for demand is 150 to 160 million tons, or between 19 and 20 kg per person in 2030. Global increases in consumption of food fish will take place predominantly in the developing countries, where population is growing and higher incomes are allowing purchase of high value fisheries items for the first time by many people. However, fish production in least developed countries where fish protein is needed to prevent malnutrition is a key element of food security in these regions and a critical area where innovative programmes are needed to increase production.

Today, fishing is the largest extractive use of wildlife in the world. The value of world total fishery production in 1999 was US$ 125 billion. World production of fish, crustaceans, mollusks and plants reached 142 million tons in 2001. Capture fisheries production, which accounted for 66 per cent of the total, was 93.7 million tons, of which inland capture was 8.7 million tons, while aquaculture production was 48.4 million tons including plants. Marine and freshwater fish are also an increasingly important recreational resource, both for active users such as anglers and for passive users such as tourists, sports divers and nature-lovers.

Predominates in capture fisheries and aquaculture production

Asia predominates in capture fisheries and aquaculture production, with China the leading nation worldwide. In 1999, over three-quarters (or 97 million tons) of the global production of fish, crustaceans and mollusks were utilized for direct human consumption. Fish utilized as raw material for the production of animal feed, *e.g.*, fish meal and fish oil, represented about one-quarter of the total fishery production in 1999; and this amount has remained relatively static for the past 15 years. Most growth in the fisheries sectors is projected to occur in developing countries, which will account for 79 per cent of food fish production in 2020.

Increased production of fish from aquaculture has occurred primarily as a result of increasing feed inputs into ponds and other production systems, thereby increasing yields per hectare by an order of magnitude compared to extensive production systems in which rearing water is fertilized only. Higher inputs mean two things to the aquaculture feed industry; more feed and higher quality feed. Currently, global feed production for farmed fish and crustaceans is approximately 13 million tons, and predictions are for feed production to increase to over 37 million tons by the end of the decade, an increase of 24 million tons. Feeds for salmonids and marine fish have always been complete feeds, *i.e.*, ones that supply all of the nutritional needs of the fish. Pond-reared fish, in contrast, obtain a significant proportion of their nutritional needs from pond biota. The degree to which feeds must supply essential nutrients to pond-reared fish increases as rearing densities increase beyond the capacity of natural foods in ponds to supply them. Fish farmers around the world have found that as they increase feed inputs, the biomass and economic yields from ponds increase as well.

Thus, great areas of low-input, pond-based aquaculture mainly in Southeast Asia and China are being converted from low-input systems to high-input systems that depend upon high quality feeds to supply an increasing proportion of nutrients used by the fish.

The effects of the aquaculture industry's growth and of changes in feed input in pond-based aquaculture have been dramatic with respect to the use of marine proteins by the aquaculture feed industry. In the mid-1980s, less than 10 per cent of annual fish meal production was used by the aquaculture feed sector. Today, that proportion is over 40 per cent. Similarly, aquaculture now uses nearly 75 per cent of annual global fish oil production, up from less than 10 per cent seventeen years ago, but this change is mainly due to the adoption of high lipid feeds by the salmon farming industry, rather than increasing the inputs into pond-based aquaculture systems.

Fish meal and oil are produced from species of fish that are not generally utilized directly for human food, *e.g.*, herring and capelin in Norway and Iceland, sand eel in Denmark, capelin in South Africa, anchovies in Peru and northern Chile, jack mackerel in central Chile, sardines in Japan, and menhaden in the USA. The rapid increase in fish meal and oil use in aquaculture feeds over the past 15 years has not resulted in over-exploitation of the stocks of fish harvested primarily to produce fish meal and oil. Production of fish meal has averaged between 6-7 million tons per year since the mid-1980s, except in El Nino years, when production has been lower. Fish oil production has averaged between 1.2-1.3 million tons. No significant changes in annual harvest or fish meal/oil production associated with increased aquaculture production or increasing intensification of aquaculture and concomitant higher feed inputs is evident. Increased fish meal/oil use in the aquaculture sector has come at the expense of other uses.

Estimates of future protein requirements for aquaculture feeds depend upon future production from various segments of the aquaculture industry and on annual production levels of fish meal. As aquaculture production expands, it will be essential to replace portions of fish meal and oil in fish feed formulations with alternative ingredients derived from alternative sources, primarily grains and oilseed products.

Because 70 per cent of the fish meal used in diets for fish and crustaceans is used to produce diets for salmonids, marine fish and shrimp, these production sectors are the focus of most research with respect to the use of alternative protein sources. Numerous studies have been conducted to evaluate the effects of replacing various percentages of fish meal in diets for these fish, and, without exception, none has successfully replaced 100 per cent of fish meal without reducing fish performance. At best, 50 per cent of fish meal in diets for salmon and trout can be replaced by soy protein concentrate, and 25-30 per cent with soybean meal. Similar findings have been reported in studies of wheat gluten meal, corn gluten meal, and rapeseed protein concentrate. For the most part, formulated fish diets are used to produce high-value fish for export. In underdeveloped countries, increased production of lower-value fish that are consumed by local populations will depend upon the use locally-available, low-cost feed ingredients that can be combined to produce prepared feeds to increase productivity of community ponds and waterways.

This is an area that requires substantial investigation and development, and is a logical target for USAID support. Another logical target for USAID efforts is the development and testing of feeds that are based upon sustainable plant-derived protein sources for use in developing countries where fish are grown both for domestic consumption and export, such as China. Partnerships with US commodity groups could be a mechanism to extend USAID efforts. It is critical that USAID become a leader in this area, given the perception that increased aquaculture production leads to higher fish meal use and greater pressure on stocks of fish that are captured to produce fish meals and oils. Fisheries products have become the most internationally traded food, as some 37 per cent (by quantity) of all fish for human consumption is traded across borders. In 1999, international trade (in live weight equivalent) represented 34 per cent of the total production. In 1999, foreign trade earnings amounted to US$ 52.9 billion. Most fishery exports were destined to developed countries, with Japan the world's leading importer, and the USA second. The US balance of trade deficit in seafood products is approximately $9 billion.

Exports of fisheries products to the developed countries have become so lucrative that nations like Argentina, a traditional, globally important exporter of meat products and livestock have turned to seafood exports as the major source of foreign exchange earnings. In 2002, exports of raw and processed fish and shellfish from Argentina surpassed meat products and livestock, with beef earning US$ 574 million in export revenue, while seafood—mostly prepared products—earned US$ 714 million. Argentina's main consumers of domestic seafood products were Spain, Brazil and the US.

2

Fish Anatomy

Brief introduction to the anatomy of fish. As with any group containing 28,000 different species there is a great deal of diversity amongst the animals we call fish. Accepting that there is a great deal of variety within the world of fish we may wish to ask ourselves what factors control the shape that a fish takes.

Many factors are involved but perhaps the most important is where the fish lives, *i.e.*, in water. Water is a very different substance than the air that we live in, it is much more supportive for one thing. This means that it is easier for fish to fly than it is for us, and they can even float, of course we call flying in water swimming, but the fact remains that because they live in water many fish can live their whole lives without ever touching the bottom, or resting on anything.

The other side of the coin of this supportiveness is that water is much harder to move through, we can run through air much faster than we can run through water. In its simplest form this means that fish don't need legs to stand on, or large wings to fly, small wings (fins) will do fine and secondly it means that if you wish to move quickly it is best to have a streamlined shape. And this of course is what we see in most fish, a streamlined shape with no legs and small wings (fins).

Like all creatures fish need to perceive the world they live in, and thus we find they have senses, sight, hearing and smell, but also, because they live in water, they can sense transverse waves generated by motion and so they have evolved a lateral line system to do this, and some have even learned to use electricity, another sense that is only really made possible because water is such a good conductor of electricity.

Within the constraints of the basic form fish have evolved further variation to accommodate different ways of living, some live on the bottom and have become dorsoventrally flatter, others hide in crevices and have become laterally flattened, some have lost their streamlined shape to facilitate a life of slow movement, and so it goes on; where you live, what you eat, who you want to hide from and how you do this all contribute to your eventual shape, both internal and external. The many forms and shapes of the fish that fill up the waters of our world are not only beautiful but practically evolved to suit a wide range of different life styles. If you are looking for information specifically on the Swim Bladder, Skeleton, the Scales, the Guts, Nerves, Blood and Respiration system please follow the links to the appropriate pages.

The Fish's Fin

Fins are essential to fish, the predecessors of legs they give a fish lift, steering capability, braking and momentum. In some species they are used to hold on to the substrate while in others they allow the fish to walk, or even to fly.

Without fins a fish would just be a fancy worm. The earliest known fishes had fins, albeit not as many as modern fish, but even the most ancient and primitive agnatha had a caudal fin. The cephalaspids had one or two dorsal fins and an anal fin as well as a pair of primitive pectoral fins. By the time of the placoderms pelvic had also evolved.

The shape and structure of a fish's fins reflect both its lifestyle and its evolution.

The ancestors of modern fish, both bony and cartilaginous, looked, in terms of their fin structure, much like modern dogfish and sharks. The evolution of the modern design, with a symmetrical tail and highly manoeuvrable body fins was only possible after the evolution of the swimbladder.

The exceptions to this design are the Birchirs, Paddlefishes and Sturgeons, which are among the most ancient and primitive of the bony fishes and still retain the larger upper lobe to the caudal fin, and some of the flying fish in which the lower lobe of the caudal fin is larger.The reasons for the asymmetry in the early fishes, and in modern cartilaginous fish is that in these fish the fins supply all, or most of the lift to the fish's body.

A fish is naturally heavier than water, even with lightweight bones other weight reducing characteristics they still have a density of around 1.076, whereas fresh water has a density of 1.0 and saltwater of about 1.026. Thus a fish will naturally sink to the sea floor if nothing stops it. If the fish is happy to live on the sea floor this is not a problem, but if it wishes to hunt nearer the surface then it has to keep making an effort to stay up. Therefore the early fish evolved an asymmetrical tail which supplies forward momentum with the top lobe and lift with the smaller bottom lobe, which, not being stiff, flexes in counterpoint to the upper lobe, and pectoral fins that work much the way the wings of a bird or an aeroplane do. Thus the tail lifts the back of the fish and the pectoral fins lift the front. Of course the fish has to keep moving, if it takes a rest it will still sink. You will probably have noticed that sharks and rays are either moving constantly, or resting on the bottom, they do not hover in the water column.

The evolution of the swimbladder, and the neutral buoyancy it gives the fish which possess it, set fish free. They could now devote the whole of the caudal fin to propulsion and develop the pectoral and pelvic fins to maximize their manoeuvrability, thus modern fish can swim backwards, sideways, come to a complete halt, hover and turn on a dime. Abilities which have greatly increased the habitats they can take advantage of and undoubtedly contributed considerably to their being the most numerous group of vertebrates on the planet.

The fins of cartilaginous fish also differ in their basic internal anatomy from those of bony fish. The fins of all fish are supported by finrays, however in cartilaginous fish these fin rays are inflexible, unbranched and unsegmented. Whereas in bony fish they are branched, flexible and segmented, except in the ancient coelocanths whose finrays are unbranched. Spines are associated with the fins of most fish. They are round in cross-section, stiff and unsegmented. They are an useful means of defence for a fish because they effectively increase its size without a great increase in living costs. Spines, when held stiff, make small fish much harder to swallow. In some species these spines are associated with poison glands which increases their effectiveness.

Functional Anatomy

The mouth exhibits a variety of fascinating adaptations for capturing, holding and sorting food, ratcheting it into the oesophagus and otherwise manipulating it prior to entry into the stomach. Only two which have possible relevance to digestion Will be discussed.

In milkfish (Chanos), the gill cavity contains epibranchial (suprabranchia) organs dorsally on each side, consisting either of simple blind sacs or elaborate, spirally-coiled ducts. The organs occur in several relatively unrelated families of lower teleosts and apparently relate to the kind of food eaten. Those fish with simple ducts all eat macro-plankton and those with the larger ducts microplankton. Although their function is unknown, concentrating the plankton has been suggested as a possibility.

The common carp provides an excellent example of non-mandibular teeth being used as the primary chewing apparatus. Pharyngeal teeth occur in the most fully developed forms of the Cyprinidae and Cobitidae, although many other groups also show some degree of abrading or triturating ability with some part of the gill bars. In carp, the lower ends of the gill bars have a well developed musculature which operates two sets of interdigitating teeth so as to grind plants into small pieces before swallowing them. The grinding presumably increases the rather small proportion of plant cells which can otherwise be successfully attached by digestive enzymes.

Many fish which chew their food have some ability to secrete mucus at the same time and place. This would have some apparent benefit when ingesting abrasive food. Although one might be tempted to equate such secretions with saliva, enzyme activity in the mucus does not appear to have been demonstrated, so the mucus is only partly comparable to saliva.

The oesophagus, in most cases, is a short, broad, muscular passageway between the mouth and the stomach. Taste buds are usually present along with additional mucus cells. Freshwater fishes are reputed to have longer (stronger?) oesophageal muscles than marine fish, presumably because of the osmoregulatory advantage to be gained by squeezing out the greatest possible amount of water from their food (*i.e.*, marine fish would be drinking seawater in addition to that ingested with their food and freshwater fish would have to excrete any excess water). The oesophagus of eels (Anguilla) is an exception to this general pattern. It is relatively long, narrow, and serves during seawater residence to dilute ingested seawater before it reaches the stomach. A possible conflict between the osmoregulatory and digestive roles of the gut in marine fish in general will be discussed later.

Fish stomachs may be classified into four general configurations. *These include*:

i. A straight stomach with an enlarged lumen, as in Esox,
ii. A U-shaped stomach with enlarged lumen as in Salmo, Coregonus, Clupea,
iii. A stomach shaped like a Y on its side, *i.e.*, the stem of the Y forms a caudally-directed caecum, as in Alosa, Anguilla, the true cods, and ocean perch,
iv. The absence of a stomach as in cyprinids, gobidids, cyprinodonts gobies, blennies, scarids and many others, some families of which only one genus lacks a stomach.

The particular advantage of any configuration seems to rest primarily with the stomach having a shape convenient for containing food in the shape in which it is ingested. Fish which eat mud or other small particles more or less continuously have need for only a small stomach, if any at all. The Y-shaped stomach, at the other extreme, seems particularly suited for holding large prey and can readily stretch posteriorly as needed with little disturbance to the attachments of mesenteries or other organs. Regardless of configuration, all stomachs probably function similarly by producing hydrochloric acid and the enzyme, pepsin.

The transport of food from the stomach into the midgut is controlled by a muscular sphincter, the pylorus. The control of the pylorus has not bean demonstrated in fish, but the best guess at this time is that it resembles that in higher vertebrates. The pylorus is developed to various degrees in different species for unknown reasons, in some species even being absent. In the latter case, the nearby muscles of the stomach wall take over this function, which may also include a grinding function by the roughened internal lining. In fish which lack a stomach, the pylorus is absent and the oesophageal sphincter serves to prevent regress of food from the intestine, *i.e.*, in fish lacking a stomach and pylorus, the midgut attaches directly to the oesophagus.

The digestive processes of the midgut have not been studied extensively, except histo-chemically, but so far as known resemble the higher vertebrates. The midgut is mildly alkaline and contains enzymes from the pancreas and the intestinal wall, as well as bile from the liver. These enzymes attack all three classes of foods – proteins, lipids, and carbohydrates – although predators such as salmonids may be largely deficient in carbohydrases. The pyloric caecae attached to the anterior part of the midgut have attracted considerable attention because of their elaborate anatomy and their taxonomic significance. Histological examination has proved them to have the same structure and enzyme content as the upper midgut. Another suggestion was that pyloric caecae might contain bacteria which produce B-vitamins as in the rodent caecum. When tested, this hypothesis had no factual basis either. Pyloric caecae apparently represent a way to increase the surface area of the midgut and nothing more. This still leaves an interesting question of how food is moved into and out of the blind sacs which are often rather lone and slim: *e.g.*, in salmonids.

The demarcation between midgut and hindgut is often minimal in terms of gross anatomy, but more readily differentiated histologically – most secretory cells are lacking in the hindgut except for mucus cells. The blood supply to the hindgut is usually comparable to that in the posterior midgut, so presumably absorption is continuing similarly as in the midgut. Formation of faeces and other hindgut functions appear to have been studied minimally, except histologically.

Peristalsis and its Control

Peristalsis consists of a travelling wave of contraction of the circular and longitudinal layers of muscle in the gut wall such that material inside the gut is moved along. The pharmacology of this system has been investigated in isolated trout intestine demonstrating that an intrinsic nerve network exists to control peristalsis; *i.e.*, cholinergic drugs stimulated and adrenergic drugs inhibited peristaltic movements. The oesophagus arid stomach are also innervated extrinsically by branches of the vagal (cranial X) nerve.

Gastric Evacuation Time and Related Studies

Many studies have been performed relating to developing an optimum feeding schedule, mostly for salmonids, but also including a number of other cultured fish. Variables considered with feeding rate and gastric evacuation time included temperature, season, activity, body size, gut capacity, satiety, and metabolic rate. A relatively consistent finding has been that gastric emptying rate declines more or less exponentially (sometimes linearly) with time. Larger meals first are often, but not always, digested at a faster rate than small meals and the amount of pepsin and acid produced was somewhat proportional to the degree of distension of the stomach. Stomach mobility often increases with the degree of stomach distension also. The appetite, digestion rate, and amount of secretions produced all decreased with decreased temperature, but the secretions also decreased if tested at temperatures in excess of the acclimation temperature. Appetite, *i.e.*, the amount of food eaten voluntarily at one time, appears to be the inverse of stomach fullness, although this does not explain the entire appetite phenomenon. Appetite continues to increase for a number of days after the stomach is empty, indicating that additional metabolic or neural mechanisms are operating. Data on gastric emptying time, digestion rate, and temperature for sockeye salmon have been shown to reflect the underlying phenomenon. Direct comparison of data on digestion among different workers is difficult, because of differences in species, food and methods used.

The total time for passage of food through the gut until the non-digestible portions of a meal are voided as faeces has not commonly been measured. Gastric emptying time and total passage time in skipjack tuna at 23-26 C was about 12 hours with the intestine being maximally filled about five hours after eating and empty after about 14 hours. Defaecation often occurred 2-3 hours after a meal, presumably being material from a previous meal. After a single meal, faeces were found 24, 48 and even 96 hours after the meal. Thus, there is considerable variation in food passage time, presumably relating to the digestibility of the food. Magnuson commented that the passage rates in skipjack tuna were at least twice as fast as known for any other fish.

The obvious importance of food passage time becomes apparent when one wishes to analyse faeces resulting from ingestion of a specific meal. If one waits to feed a test meal until the gut is completely empty, then the digestion processes observed will be typical only of starved fish. If one feeds the test meal as part of a regular feeding programme, then the problem is to mark the food for appropriate faecal analysis. Thus the problem is not as simple as it might appear at first.

Digestion and Absorption

Digestion is the process by which ingested materials are reduced to molecules of small enough size or other appropriate characteristics for absorption, *i.e.*, passage through the gut wall into the blood stream. This generally means that proteins are hydrolyzed to amino acids or to polypeptide chains of a few amino acids, digestible carbohydrates to simple sugars, and lipids to fatty acids and glycerol. Materials not absorbed are by definition indigestible and are eventually voided as faeces. Digestibility ranges from 100 per cent for glucose to as little as 5 per cent for raw starch or 5-15 per cent for plant material containing mostly cellulose (plant fibre). Digestibility of most natural proteins and lipids ranges from 80 to 90 per cent.

Digestion is a progressive process, beginning in the stomach and possibly not ending until food leaves the rectum as faeces. Most studies of digestion simply compare the protein, lipid and carbohydrate content of the faeces with that of the feed. A study on digestion in channel catfish by Smith and Lovell showed continuing digestion (and absorption) of protein during passage through each part of the gut. The comparison of faeces collected from the rectum and from the water also points out the hazard of incomplete recovery of faecal matter being likely when collection is done from outside the gut. Most of the protein digestion occurred in the stomach, but also continued in the intestine.

Table. Apparent Digestibility of Protein by Channel Catfish

Feed	Stomach	Upper Intestine	Lower Intestine	Rectum	Trough
20% protein	61.6	65.4	75.0	80.9	96.7
40% protein	61.4	72.2	86.5	96.6	98.3

Temperature and pH play major roles in determining the effectiveness of digestive enzymes as a whole. Although most enzyme production decreases at temperatures above or below acclimation temperature, most enzyme activity (for a given amount of enzyme) increases in proportion to the temperature over a wide range of temperatures.

In general, enzyme reaction rates continue to increase at higher temperatures, even though the temperatures increase beyond the lethal temperatures for the species, until the enzymes begin to denature around 50-60°C. On the other hand, enzymes have limited ranges of pH over which they function, often as little as 2 pH units. Data for channel catfish are probably representative of many teleosts. Acid concentrations (pH) in the stomach ranged from 2 to 4, then became alkaline (pH = 7-9) immediately below the pylorus, decreased slightly to a maximum of 8.6 in the upper intestine, and finally neared neutrality in the hindgut. Fish having no stomach have no acid phase in digestion.

The site of secretion in teleost stomachs appears to be a single kind of cell which produces both HCl and enzyme(s). This contrasts with mammals where two types of cells occur, one for acid and one for enzymes. The production of acid in teleosts is presumably the same as in mammals – NaCl and H_2CO_3 react to produce $NaHCO_3$ and HCl, with the blood being the source of both input materials, which are later mostly reabsorbed in the intestine. One possible explanation for the loss of stomachs in some species of fish is that they live in a chloride-poor environment and that providing large amounts of chloride ion for operating a stomach is bioenergetically disadvantageous. In addition to acid and enzymes, the stomach wall also secretes mucus to protect the stomach from being digested. As long as the rate of mucus production exceeds the rate at which it is washed and digested away, the gut wall is protected from being digested. When mucus production slows or fails, *e.g.* during gut stasis, during stressful conditions, or post mortem, the gut wall can be eroded or even perforated by the gut's own digestive enzymes.

Two sites produce enzymes in the midgut - the pancreas and the intestinal wall. The intestinal wall is folded or ridged in simple patterns which can be species specific. Secretory cells for both mucus and all three classes of enzymes develop in the depths of the folds, migrate to tops of the ridges (closest to the gut lumen), and then discharge their products. The pancreatic cells produce enzymes and an alkaline solution which are delivered to the upper midgut through the common bile duct. The control of pancreatic secretions (and the pyloric sphincter) in fish is probably the same as in mammals, but there is no information on teleosts yet.

The physical state of food passing through the gut varies with species and type of food. Fish, such as salmonids, which eat relatively large prey, reduce the prey in size layer by layer. Gastric digestion proceeds in a layer of mucus, acid, and enzyme wherever the stomach wall contacts the food. Food appears liquified only in the midgut and solidifies somewhat again during formation of faeces.

Pellets of commercial feed seem to be treated similarly, *i.e.*, pellets get smaller and smaller in size with time, although stomachs of some recently-fed salmonids have been found to contain moderate amounts of liquified pellets. Stomachs of juvenile Pacific salmon captured in the open sea contained a thick slurry of pieces of amphi-pods in various stages of solubilization. Fish whose food contains high levels of indigestible ballast, *e.g.*, common carp feeding on a mixture of mud and plants, probably show minimal change in the appearance or volume of their food while it passes through the gut. Microphagous fish, such as the milkfish (Chanos) whose food starts out as a suspension of fine particles, probably also keep it in much the same form all the way through the gut. In general: there seems not to be the same degree of liquifaction of food in fish as is commonly described for mammals.

Absorption of soluble food could begin in the stomach – it occurs in mammals, but has not been investigated in fish – but takes places predominantly in the midgut and probably to some degree in the hindgut. The sites and mechanisms of absorption are largely unstudied, except histologically. Several histologists have identified fat droplets in intestinal epithelial cells following a lipid-rich meal. Increased numbers of leucocytes in general circulation following a meal by the sea bream and increased number of fat droplets in them have been described.

It was hypothesized that leucocytes entered the gut lumen, absorbed lipid droplets, and then returned to the blood stream. It is clear that the mammalian type of villi with their lymph duct (lacteal) inside are absent in fish, although there is some folding and ridging of the gut wall to increase surface area. Lacteals serve as a primary uptake route in mammals for uptake of droplets of emulsified lipids (chylomicra). Teleost fish have a lymphatic system which includes extensions into the gut wall, but its role in lipid uptake is unknown. Absorption of amino acids, peptides, and simple carbohydrates have been little studied, but presumably they diffuse through or are transported across the gut epithelium into the blood stream. What light microscopists identified as a brush border on the surface of the epithelial cells facing the gut lumen, has now been clarified with electron-microscopy as microvilli; *i.e.*, subcellular, finger like projections of the cell membrane whose greatly increased surface area is probably involved in absorption.

Specific Dynamic Action (SDA)

Digested food, particularly proteins, is not fully available to a fish even after it has been absorbed into the blood stream. Amino acids, if used for building new tissue, could be used as absorbed. If amino acids are to be oxidized for energy, however, deamination (removal of the amino group) must occur first – a reaction which requires input of energy. This process, known as specific dynamic action (SDA), can be measured externally in fish as an increase in oxygen consumption beginning soon after ingestion of food followed by an increase in ammonia excretion.

The proportion of amino acids which get deaminated varies with the food and the fish's circumstances. Fish which are not growing because of low temperature or have their ration at maintenance level or below, would deaminate most or all of their amino acids. Fish kept at high rearing temperatures or at high activity levels and therefore having very high metabolic rates would do likewise. On the other hand, fish having rapid growth and high protein intake would deaminate a relatively small proportion of their digested protein, although the absolute quantity of amino acids deaminated could still be large enough to produce a relatively large SDA. The energy for deamination need not necessarily come from amino acids, but will be preferentially taken from carbohydrate or lipid, if available. Thus, salmonid aquaculturists long ago discovered this "protein-sparing" action of limited amounts of inexpensive carbohydrate in the diet as a way of reducing the cost of feed and still achieving a desired level of growth. The protein-sparing action of lipids appears to have been minimally investigated. One can thus minimize SDA costs, but not avoid them completely.

Interrelationship between Osmoregulation and Digestion

Researchers studying osmoregulation and researchers studying digestion have rarely considered each other's data. Marine fishes drink significant amounts of seawater, a relatively-well buffered solution having a pH of about 8.5, while gastric digestion requires a pH of 4 or lower in most fish.

The amount of HCl required just to acidify the seawater would be substantial, that is, if the entire stomach gets flooded with seawater. There are several likely alternatives, however. In fish with Y-shaped stomachs, the seawater could travel directly from the oesophagus to the pylorus, and traverse only a small fraction of the stomach surface. If, at the same time, digestion functioned primarily as contact digestion, then it could be largely separated from osmoregulation. On the other hand, marine salmon stomachs have been found to be filled with a liquid slurry which would prevent such separation. In such cases, alternation of digestion and seawater drinking might be possible, although fish whose stomachs seemed continuously filled, and therefore would have no time for drinking, have also been observed.

The pH of seawater should cause little or no problem with intestinal digestion. Too high a salt content in the intestine might exceed the operational range of some enzymes and thus reduce the rate of digestion. However, one of the functions of the stomach (and in eels, the oesophagus) in osmoregulation is to dilute the incoming seawater until it is approximately equal to the osmolarity of blood, thus protecting the intestine.

The final osmoregulatory product of the gut is a rectal fluid composed of magnesium and other divalent ions having about the same total concentration as blood. Preliminary data from scale loss studies indicated that death occurred from toxic levels of magnesium in the blood. A possible cause of the high magnesium is that gut peristalsis stopped, leaving the rectal fluid to accumulate and the magnesium ions to be reabsorbed instead of being excreted. Thus, digestion and osmoregulation are so inter-related that problems in one system could disrupt the functions of the other. Exactly how fish normally avoid such problems is largely unknown.

Characteristics

The ability of any organism to digest a given substance rests predominantly on whether the appropriate enzyme is present or not and then whether the required conditions for operation of that enzyme exist or not. The following describes the enzymes and their requisite conditions according to their location.

Digestion in the Mouth and Oesophagus

The hard surfaces of the mouths of most teleost; fishes would not lead one to expect any kind of secretion. However, many fish which chew with pharyngeal teeth or similar structures also produce mucus while chewing. Tests of this mucus in a few species for enzyme activity have so far yielded negative results. Likewise, oesophageal mucus cells, when examined histologically, showed no sign of containing any enzymatic granules, although there are reports of gastric-like secretory cells in the posterior oesophagus of a few fish.

Digestion in the Stomach

Pepsin is the predominant gastric enzyme of all vertebrates, including fish.

Optimal pH for maximal proteolytic activity has been reported for several species, as follows:

- pH 2 – Pike, plaice
- pH 3-4 – Ictalurus
- pH 1.3, pH 2.5-3.5 – Salmon, probably similar for tuna

Peptic activity has been shown in a number of cultures and commercial species including Anguilla japonica, Tilapiamossambica, Pleuronecthys, both Salmo and Oncorhynchus species, Ictalurus, Micropterus, Lepomis and Perca. The presence of pepsin is so universal in vertebrates having stomachs that its presence can be presumed in fish for which no data is available.

The histochemistry of gastric secretion has been little studied in fish, although there is agreement on the presence of only one type of secretory cell in fish which stains positively for indicators of pepsinogen (pepsin precursor) cells.

There is some question whether there may be more than one pepsin present in some fish, but no chromatographic or other tests have been done to investigate this. Several attempts have been made to identify acid-secreting cells, but results were either negative or confusing. Other gastric enzymes have been proposed, but not firmly identified.

Chitinolytic activity with an optimum at pH 4.5 was claimed for the stomach of Salmo irideus, but in most cases is probably from exogenous sources. If fish are like higher vertebrates, then the stomach wall also produces the hormone gastrin which stimulates gastric secretion. A lipase may also be present.

Digestion in the Midgut and Pyloric Caecae

There are two sources of enzymes for the midgut - the pancreas and the secretory cells in the gut wall - with the pancreas perhaps secreting the greater variety and quantities of enzymes in fish. Because of the variety of enzymes present in different species, there have been some attempts to correlate enzyme activities with diet. However, these enzyme studies are fragmentary and histochemical tests are too general. Much remains to be learned about intestinal digestion in fish.

Trypsin appears to be the predominant protease in the midgut. Since the enzyme appears not to have been isolated, most authors have just tested for proteolytic activity over the pH range of 7 to 11 and reported their results as tryptic activity. The diffuse nature of the pancreas in most cases has limited many researchers to making relatively crude extracts from mixed tissues, hampering localization of the enzyme. Tryptic activity has been found in four stomachless species in Japan: Seriola,two basses and a puffer.

Since these fish lack pepsin, some such kind of protease in the intestine would be the primary means of protein digestion. Tryptic activity was found in extracts of both the pancreas of perch and Tilapia and in intestinal extracts of Tilapia, all having a pH optimum of 8.0-8.2. Proteolytic activity has been identified in the pyloric caecae and intestine of rainbow trout. In grass carp, tryptic activity was stronger in the intestine than in the pancreas. In a mixture of pancreatic and pyloric caecae tissue from chinook salmon, casein was digested maximally at pH 9.

Tryptic activity has also been demonstrated in extracts of liver of Several species, probably because in fish having a diffuse pancreas, pancreatic tissue extends into the liver, around the portal veins, and around the gall bladder.

In several of the cases above, when extracts of pancreas were mixed with extracts of intestine, the tryptic activity increased ten-fold or more, suggesting the presence in fish of the enzyme enterokinase in the intestinal wall which activates in mammals the pancreatic trypsin as it reaches the intestine. Additional pancreatic enzymes are involved in midgut digestion, many of them yet to be discovered. For example, Japanese workers are studying the occurrence and characteristics of a pancreatic collagenase in several Japanese fishes. There have also been several reports of chitinolytic activity in some fish which eat crustaceans predominantly. This could also have resulted from bacterial activity.

The occurrence of at least one lipase may be assumed in all fishes and has been demonstrated for a number of species. In carp and killifish extracts of intestine showed lipolytic activity. In goldfish, lipase activity occurred in extracts of a mixture of liver and pancreas and in the intestinal contents. Esterase (another lipase) activity has been found in the liver, spleen, bile, intestine, pyloric caecae and stomach of rainbow trout. Use of radioisotope-labeled lipids in cod suggested that the cod's lipase acted in the same manner as mammalian pancreatic lipase, although it was not considered more than a suggestion that fish lipase is of pancreatic origin. Regardless of origin, some kind of lipase is essential to fish because fatty acids are essential dietary components for fish.

Carbohydrases have perhaps excited the most interest of all the enzymes, particularly because salmonids do not handle the large carbohydrate molecules very well, and many workers wanted to determine the reason.

Further, because there are several carbohydrases, the possibility that different enzyme combinations might show adaptations to different diets also intrigued some investigators. Also, herbivorous fish might be expected to have more carbohydrase activity and less tryptic activity than carnivores or omnivores.

Amylase is a widespread starch-digesting enzyme which occurs in human saliva and in pancreatic secretions into the small intestine. Amylase activity has been found in goldfish and bluegill sunfish in extracts of mixed liver and pancreas, oesophagus (contamination from regurgitated food suggested) and intestine, but not in large-mouth bass. Similar activity has been seen as well in rainbow trout, perch, Tilapia, Pacific salmon, cod, common carp, eel, and flounder. In fish with a diffuse pancreas there may be no pancreatic duct and so amylase activity appears in the bile. In mackerel. Scomber spp., which have a compact pancreas, the bile had no amylase activity.

Other carbohydrases identified included glucosidases (rainbow trout, chum salmon, common carp), maltase (common carp, red sea bream, Archosargus, marine ayu, Plecoglossidae), and sucrase, lactase, melibiase, and cellobiase, all of the latter in common carp.

The hypothesis that carnivores might be deficient in one or more carbohydrases is largely disproved by the widespread presence of amylase in salmonids and other predators and by the presence of maltase in sea bream and ayu. The apparently larger diversity of carbohydrases in common carp than in other fish seems mostly a lack of information about fish other than carp.

The question of whether dietary differences influence the kind of enzymes present must remain open but the evidence so far remains largely negative. However, there seems to be some evidence to show that the amounts of various enzymes may relate to the diet. Data in Table suggest that herbivores have de-emphasized the production of proteases compared to the carnivores and the reverse for carbohydrases.

Similarly, in studies of Trachurus, Scomber, Mullus, Mugil, and Pleuronectes, the predatory species, Trachurus andScomber had the highest proteolytic and lipolytic activities, while the planktivore, Mugil, had the lowest proteolytic and the highest amylolytic activities. Also, stomachless fish (which lack pepsin) are usually herbivores or omnivores, while carnivorous fish have true stomachs with peptic digestion. On the other hand, differences in proteolytic activity betweenTilapia and Perca were small, and some other investigations of a variety of species failed to find any species differences. Apparently, where fish are somewhat specialized in their diets, differences in their enzyme activities are apparent. Many fish, however, remain non-specialized and have diversified diets and enzymes.

Role of Bile, Gall Bladder and Liver in Digestion

The functions of bile have scarcely been studied in fish, but presumably resemble those in higher vertebrates. In mammals bile is composed mainly of bilirubin and biliverdin, which are breakdown products of haemoglobin, and is produced continuously. These salts act like detergents and serve to emulsify lipids, thus making lipids more accessible to enzymes because of the increased surface area, allowing some lipids to be absorbed undigested as micro-droplets. In mammals, about 80 per cent of the bile is recycled through the liver and gall bladder.

There are a few studies in fish which suggest that bile serves similar functions in fish. Several histologists have histochemically identified micro-droplets of lipid in midgut epithelium of fishes. That the gall bladder in fish reabsorbs water as in mammals has been confirmed. That bile is produced continuously in fish is suggested by the presence of green mucus in the lumen of the atrophied gut of spawning salmon. There appear to be no studies in fish of gall bladder contraction or other mechanisms controlling the release of bile during digestion. An observation of salmon having impacted gall bladders seemed related to diet because the gall bladders returned to normal when their dry pellet diet was changed to a moist pellet. Fish having impacted (and presumably non-contractile) gall bladders were normal otherwise and were indistinguishable in appearance and growth rates from fish in the same population with normal gall bladders.

Anatomists have tried for many years to correlate the shape of the liver and the position of the gall bladder in the liver with some of its functions. The basic functions of the liver in processing the foods which have been digested and absorbed are entirely cellular and molecular in scope. Thus, there is no functional requirement for shape at any level above the cellular level; *i.e.*, livers basically could be of any shape. On the other hand, some restrictions are created by its position in the circulatory system between the gut and the heart, and the necessary interdigitation of the portal and hepatic veins, hepatic arteries, and bile ducts, all of which must serve essentially every cell of the liver. In common carp, the liver seems to have no shape of its own and simply fills every available space between the loops of the intestine. On the other hand, many fish (*e.g.*, salmonids) have distinctive shape and colour to their livers. Changes in normal size and shape can indicate dietary or other problems. For example, a large, yellowish liver, often with white blotches suggests fatty degeneration of the liver caused by too much starch or by using saturated (mammalian) fats in the diet.

Anatomy and Physiology of Fish

The Skeleton

Being vertebrates, fish have a vertebral column - the backbone - and a cranium covering the brain. The backbone runs from the head to the tail fin and is composed of segments (vertebrae). These vertebrae are extended dorsally to form neural spines, and in the trunk region they have lateral processes that bear ribs. The ribs are cartilaginous or bony structures in the connective tissue (myocommata) between the muscle segments (myotomes). Usually, there is also a corresponding number of false ribs or "pin bones" extending more or less horizontally into the muscle tissue. These bones cause a great deal of trouble when fish are being filleted or otherwise prepared for food.

Muscle Anatomy and Function

The anatomy of fish muscle is different from the anatomy of terrestrial mammals, in that the fish lacks the tendinous system connecting muscle bundles to the skeleton of the animal. Instead, fish has muscle cells running in parallel and connected to sheaths of connective tissue (myocommata), which are anchored to the skeleton and the skin. The bundles of parallel muscle cells are called myotomes. All muscle cells extend the full length between two myocommata, and run parallel with the longitudinal direction of the fish. The muscle mass on each side of the fish makes up the fillet, of which the upper part is termed the dorsal muscle and the lower part the ventral muscle.

The fillet is heterogenous in that the length of the muscle cells vary from the head end (anterior) to the tail end (posterior). The longest muscle cells in cod are found at about the twelfth myotome counting from the head, with an average length around 10 mm in a fish that is 60 cm long. The diameter of the cells also vary, being widest in the ventral part of the fillet. The myocommata run in an oblique, almost "plow-like" pattern perpendicular to the long axis of the fish, from the skin to the spine. This anatomy is ideally suited for the flexing muscle movements necessary for propelling the fish through the water.

As in mammals, the muscle tissue of fish is composed of striated muscle. The functional unit, *i.e.*, the muscle cell, consists of sarcoplasma containing nuclei, glycogen grains, mitochondria, etc., and a number (up to 1 000) of myofibrils. The cell is surrounded by a sheath of connective tissue called the sarcolemma. The myofibrils contain the contractile proteins, actin and myosin. These proteins or filaments are arranged in a characteristic alternating system making the muscle appear striated upon microscopic examination. Most fish muscle tissue is white but, depending on the species, many fish will have a certain amount of dark tissue of a brown or reddish colour. The dark muscle is located just under the skin along the side of the body.

The proportion of dark to light muscle varies with the activity of the fish. In pelagic fish, *i.e.*, species such as herring and mackerel which swim more or less continuously, up to 48 per cent of the body weight may consist of dark muscle. In demersal fish, *i.e.*, species which feed on the bottom and only move periodically, the amount of dark muscle is very small. There are many differences in the chemical composition of the two muscle types, some of the more noteworthy being higher levels of lipids and myoglobin in the dark muscle.

From a technological point of view, the high lipid content of dark muscle is important because of problems with rancidity. The reddish meat colour found in salmon and sea trout does not originate from myoglobin but is due to the red carotenoid, astaxanthin. The function of this pigment has not been clearly established, but it has been proposed that the carotenoid may play a role as an antioxidant. Further, the accumulation in the muscle may function as a depot for pigment needed at the time of spawning when the male develops a strong red colour in the skin and the female transport carotenoids into the eggs. The latter seems to depend heavily on the amount of carotenoids for proper development after fertilisation. It is clearly seen that the muscle colour of salmonids fades at the time of spawning. The fish cannot synthesise astaxanthin and is thus dependent on ingestion of the pigment through the feed. Some salmonids live in waters where the natural prey does not contain much carotenoid, *e.g.*, in the Baltic Sea, thus resulting in a muscle colour less red than salmonids from other waters.

This may be taken as an indication that the proposed physiological function of astaxanthin in salmonids explained may be less important. In salmon aquaculture, astaxanthin is included in the feed, as the red colour of the flesh is one of the most important quality criteria for this species. Muscle contraction starts when a nervous impulse sets off a release of Ca $^{++}$ from the sarcoplasmic reticulum to the myofibrils. When the Ca $^{++}$ concentration increases at the active enzyme site on the myosin filament, the enzyme ATP-ase is activated.

This ATP-ase splits the ATP found between the actin and myosin filaments, causing a release of energy. Most of this energy is used as contractile energy making the actin filaments slide in between the myosin filaments in a telescopic fashion, thereby contracting the muscle fibre. When the reaction is reversed (*i.e.*, when the Ca $^{++}$ is pumped back, the contractile ATP-ase activity stops and the filaments are allowed to slip passively past each other), the muscle is relaxed. The energy source for ATP generation in the light muscle is glycogen, whereas the dark muscle may also use lipids.

A major difference is, further, that the dark muscle contains much more mitochondria than light muscle, thus enabling the dark muscle to operate an extensive aerobic energy metabolism resulting in CO_2 and H_2O as the end products.

The light muscle, mostly generating energy by the anaerobic metabolism, accumulates lactic acid which has to be transported to the liver for further metabolisation. In addition, the dark muscle is reported to possess functions similar to those are found in the liver.

The different metabolic patterns found in the two muscle types makes the light muscle excellently fitted for strong, short muscle bursts, whereas the dark muscle is designed for continual, although not so strong muscle movements. Post mortem the biochemical and physiological regulatory functions operating *in vivo* ceases, and the energy resources in the muscle are depleted. When the level of ATP reaches its minimum, myosin and actin are interconnected irreversibly, resulting in rigour mortis.

The Cardiovascular System

The cardiovascular system is of considerable interest to the fish technologist since it is important in some species to bleed the fish (*i.e.*, remove most of the blood) after capture. The fish heart is constructed for single circulation. In bony fish it consists of two consecutive chambers pumping venous blood towards the gills via the ventral aorta. After being aerated in the gills, the arterial blood is collected in the dorsal aorta running just beneath the vertebral column and from here it is dispersed into the different tissues via the capillaries.

The venous blood returns to the heart, flowing in veins of increasingly larger size (the biggest is the dorsal vein which is also located beneath the vertebral column). The veins all gather into one blood vessel before entering the heart. The total volume of the blood in fish ranges from 1.5 to 3.0 per cent of the body weight. Most of it is located in the internal organs while the muscular tissues, constituting two- thirds of the body weight, contain only 20 per cent of the blood volume. This distribution is not changed during exercise since the light muscle in particular is not very vascularised.

During blood circulation the blood pressure drops from around 30 mg Hg in the ventral aorta to 0 when entering the heart. After the blood has passed through the gills, the blood pressure derived from the pumping activity of the heart is already greatly decreased. Muscle contractions are important in pumping the blood back to the heart and counterflow is prevented by a system of paired valves inside the veins. Clearly, the single circulation of fish is fundamentally different from the system in mammals, where the blood passes through the heart twice and is propelled out into the body under high pressure due to the contractions of the heart.

In fish, the heart does not play an important role in the transportation of blood from the capillaries back to the heart. This has been confirmed in an experiment where the impact of different bleeding procedures on the colour of cod fillets was examined. No difference could be found regardless of whether the fish had been bled by means of cutting the throat in front of or behind the heart before gutting, or had not been cut at all before slaughter.

In some fisheries, bleeding of the fish is very important as a uniform white fillet is desirable. In order to obtain this, a number of countries have recommended that fish are bled for a period (15-20 min) prior to being gutted. This means that throat cutting and gutting must be carried out in two separate operations and that special arrangements (bleeding tanks) must be provided on deck. This complicates the working process (two operations instead of one), time-consuming for the fishermen and increases the time-lag before the fish is chilled. Furthermore it requires extra space on an otherwise crowded working deck. Several researchers have questioned the necessity of handling the fish in a two-step procedure involving a special bleeding period.

There seems to be general agreement about the following:

- bleeding is more affected by time onboard prior to bleeding/gutting than by the actual bleeding/gutting procedure.
- best bleeding is obtained if live fish are handled, but it is of major importance to cut the fish before it enters rigour mortis since it is the muscle contractions that force the blood out of the tissues.

Disagreement exists as to the cutting method. Huss and Asenjo found best bleeding if a deep throat cut including the dorsal aorta was applied, but this was not confirmed in the work. The latter also recommended to include a bleeding period (two-step procedure) when live fish were handled (fishing with pound net, trap, seine, longline or jigging), while found that the quality of dead cod (4 h after being brought onboard) was slightly improved using the two-step procedure.

However, it should be pointed out that the effect of bleeding should also be weighted against the advantages of having a fast and effective handling procedure resulting in rapid chilling of the catch.

Discolouration of the fillet may also be a result of rough handling during catch and catch handling while the fish is still alive. Physical mishandling in the net (long trawling time, very large catches) or on the deck (fishermen stepping on the fish or throwing boxes, containers and other items on top of the fish) may cause bruises, rupture of blood vessels and blood oozing into the muscle tissue (haematoma). Heavy pressure on dead fish, when the blood is clotted (*e.g.*, overloading of fish boxes) does not cause discolouration, but the fish may suffer a serious weight loss.

Other Organs

Among the other organs, only the roe and liver play a major role as foodstuffs. Their size depends on the fish species and varies with life cycle, feed intake and season. In cod the weight of the roe varies from a few per cent up to 27 per cent of the body weight and the weight of the liver ranges from 1 to 4.5 per cent. Likewise, the composition can change and the oil content of the liver vary from 15 to 75 per cent, with the highest values being found during autumn.

Growth and Reproduction

During growth it is the size of each muscle cell that increases rather than the number of muscle cells. Also, the proportion of connective tissue increases with age. Most fish become sexually mature when they reach a size characteristic of the species and is this not necessarily directly correlated with age. In general, this critical size is reached earlier in males than in females.

As the growth rate decreases after the fish has reached maturity, it is therefore often an economic advantage to rear female fish in aquaculture. Every year mature fish use energy to build up the gonads (the roe and milk). This gonadal development causes a depletion of the protein and lipid reserves of the fish since it takes place during a period of low or no food intake.

In North Sea cod it was found that prior to spawning the water content of the muscle increases and the protein content decreases. In extreme cases the water content of very large cod can attain 87 per cent of the body weight prior to spawning.

The length of the spawning season varies greatly between species. Most species have a marked seasonal periodicity, while some have ripe ovaries for nearly the whole year. The depletion of the reserves of the fish during gonadal development can be extremely severe, especially if reproduction is combined with migration to the breeding grounds. Some species, *e.g.*, Pacific salmon (Oncorhynchus spp.), eel (Anguilla anguilla) and others, manage to migrate only once, after which they degenerate and die. This is partly because these species do not eat during migration so that, in the case of a salmon, it can lose up to 92 per cent of its lipid, 72 per cent of its protein and 63 per cent of its ash content during migration and reproduction. On the other hand, other fish species are capable of reconstituting themselves completely after spawning for several years.

The North Sea cod lives for about eight years before spawning causes its death, and other species can live even longer. In former times, 25-year-old herring (Clupea harengus) were not unusual in the Norwegian Sea, and plaice (Pleuronectes platessa) up to 35 years old have been found. One of the oldest fish reported was a sturgeon (Acipenser sturio) from Lake Winnebago in Wisconsin. According to the number of rings in the otolith, it was over 100 years old.

The Circulatory System in Fish

As in other vertebrates the circulatory system of fish is comprised of both static and dynamic components. The dynamic part is the blood with all its constituent parts that flows continuously around the fish's body. The static parts are the heart, the veins and arteries leading to and from it and the capillaries that connect them. Fish have less blood per gram of body weight than mammals, normally between 3 per cent and 8 per cent of a fish's body weight is blood, however in the hagfish and lampreys it is greater, for them between 8 per cent and 20 per cent of the animal's body weight.

The blood of fish consists of plasma and the various cells that move around the body in it, erythrocytes and leukocytes. Plasma is basically water with a variety of ions (Na+, Ca_2+, K+) and small organic molecules such as urea, sugars and fatty acids dissolved in it. In sharks and rays blood cells are created in three different organs, the spleen, the epigonal organ which surrounds the gonads and the Leydig organ which is found in the throat near the oesophagus. However in teleost fish blood cells are normally only produced in the spleen and the kidneys.

Blood cells, the biological boats that sail and trade in the biological sea of blood come in two sorts. Erythrocytes and Leucocytes. Erythrocytes are by far the most common cells found in the blood plasma and their job is to move gases around the body. they collect oxygen at the gills and carry it to all the cells of the body, and from these same cells they collect carbon dioxide which they release into the external waters through the gills. The molecule that allows the erythrocytes to do their job, and which makes them red, is called haemoglobin, or hemoglobin, and erythrocytes are full of it. The haemoglobin of hagfish and lampreys is called monomeric and consists of a single large haem-molecule with a molecular weight of around 17,000 daltons.

Most other fish have tetrameric haemoglobin which is built up from from four different protein chains called alpha and beta chains, there are two of each in a single haemoglobin molecule. This type of haemoglobin is very similar to ours and has a molecular weight of around 65,000 daltons.

Leucocytes are a mixed collection of cells that are united in being much less common that erythrocytes and in not being red because they do not contain haemoglobin. The five most important types are Granulocytes, Lymphocytes, Monocytes, Thrombocytes and Non-specific Cytotoxic Cells.

- *Lymphocytes:* Are 4.5 to 12 microns in diametre and their job is in defence in detecting and marking foreign particles, specifically they mediate antibody activity, antibodies are small molecules that find and bind to foreign materials so that Monocytes can find them.
- *Thrombocytes:* Are important in blood clotting, thus they are important in conservation of resources as they prevent blood loss in case of injury.
- *Monocytes:* Are sometimes called Macrophages, they are important in defence, their role is to eat (phagocytise) anything they come across in the blood that might harm the fish such as bacteria, or parasite larvae.
- *Granulocytes:* Are also active in defence, they specifically attack bacteria, but also seem to have a role in controlling stress, however this is still being researched.

- *Non-specific Cytotoxic Cells:* These are another defence mechanism, this time against tumors and protozoan parasites.

There are always exceptions, this is part of the beauty of life on earth. The Icefish of the Antarctic such as Pseudochaenichthys georgianus have no haemoglobin in their blood and rely on gases dissolved in the plasma, this helps them become invisible, or at least harder to see. They can survive like this because 1) they live in very cold water, cold water has a higher content of dissolved oxygen than warm water, 2) they live fairly inactive lives so their oxygen requirements are low, their food is plentiful and easy to catch.

At the centre of the static, comparatively, parts of the circulatory system is the heart which is normally situated below the pharynx and immediately behind the gills. A fish's heart has four chambers, but unlike us the chambers of their heart are not all muscular, and hey are not so built into a single organ, rather they are located one behind the other, this is particularly evident in the sharks and rays.

Sinus venosus. The first chamber is called the sinus venosus, it is the preliminary collecting chamber. In teleosts it is filled from two major veins called the hepatic veins and the left and right branches of the Curvierian ducts which in turn collect blood from the paired (left and right) lateral veins the inferior jugulars, the anterior cardinals and the posterior cardinals. However in the elasmobranchs only one hepatic vein leads into it.

Atrium. From the sinus venosus the blood flows into the atrium. the atrium is the largest of the chambers and only weakly muscular. It pushes the blood, with weak contractions in the ventricle.

Ventricle. The ventricle is the only well muscled chamber, nearly as large as the atrium it is th work horse of the heart, its contractions drive the blood around the body.

Bulbus arteriosus. The last chamber of the fish heart is called the bulbus arteriosus in the teleosts, but the cornus arteriosus in the elasmobranchs. The difference between these chambers is that the cornus arteriosus of sharks and rays contains many valves while the bulbus arteriosus of bony fish contains none. Both are alike in being primarily elastic and work to reduce the pulsed nature of the blood leaving the ventricle giving it a more even, constant flow.

When you feel your pulse, or measure a fish's pulse there is a beat and a space, beat - space - beat - space, beat - space, and so on. The beat is when the blood is being pumped out of the heart, which is when, in fish the ventricle, is contracting, the space is when the ventricle relaxes and is filled with blood again. (note that a one-way valve at the exit of the ventricle prevents the blood which was just pumped out flowing back into the ventricle). Scientists call the emptying of the heart the systole part of the pulse and the filling of the heart the diastole part of the pulse. Systole = beat and diastole = space, together they make one pulse.

Fish hearts have become more efficient through the ages, gram for gram the heart of a modern teleost fish is 2 to 5 times more powerful than that of sharks and rays, and 10 to 25 times more powerful than the heart of a hagfish. From the arteriosus the blood enters the ventral artery which takes it to the gills where it loses CO_2 and gains O_2 before heading out to the body through the dorsal artery. Various branch arteries supply the muscles and different organs of the body through a network of ever finer capillaries. Further capillaries take the blood, now low in O_2 and higher in CO_2 into a series of veins that ultimately join up with the major veins, the hepatic veins, lateral veins, inferior jugulars, anterior cardinals and posterior cardinals.

Not all th plasma of the blood returns through the capillaries, some of it mixes in with the fluid that surrounds the cells of every tissue and is eventually drained away into the lymphatic vessels. This secondary circulation of lymph vessels eventually empties into the main blood veins. However because it left the main circulatory system it is unaffected by the beating of the heart and its flow is slow. In some fish it is helped along by small lymph hearts, muscular valved tubes which help express the blood into the veins. Lymphatic hearts can be found in various places in those species that have them, in the Eel (Anguila anguila) has one lymphatic heart situated in its tail.

The Brain and Nervous System of Fish

Being highly complex life forms fish need a brain and a nervous system to control their body's actions. The nervous system of fish, much like ours, is composed of a central co-ordinating brain, a spinal cord and many nerves. The Brain: Generally speaking fish have small brains in relationship to their overall body weight. Elasmobranchs (Sharks and Rays) in general have a slightly larger brain for the same body mass as Teleosts (Bony Fish), however there is great variety within the teleosts scientists have learned something quite surprising about the Elephantnose Fish (Gnathonemus petersii). Lurking in the vegetation and then aggressively attacking anything small that comes near doesn't seem to require much intelligence. A Pike, (Essox lucius) has a brain that amounts to only 0.077 per cent of its total body mass. Which means that when you catch a 5 kilogram Pike you have out maneuvered a fish with a brain weighing 3.85 grams or one seventh of an ounce.

In comparison a Burbot (Lota lota) is a little smarter because it has a brain that is 0.13 per cent of its body weight meaning a 5 kilo specimen would have a 6.5 gram brain, still its is a pretty dumb fish. Most fish have brains that are less than 1 per cent of their body weight, but not all.

At the other end of the continuum of fish intelligence the electric Elephantnose fish, with a brain that is 3.1 per cent of its body weight it is a lot smarter than the Pike, if it grew to weigh 5 kilos its brain would weigh 155 grams, which would make it 40 times as large as that of a Pike of similar weight. To look at it another way a human being has, on average, a brain that weighs in at around 2.3 per cent of body weight. A five kilo human would have a brain of only 115 grams, which is smaller than that of the Elephantnose Fish.

Does this mean Elephantnose fish are smarter than people? No. Intelligence is more than just brain mass or even brain to body ratio, while large brains do make it easier to be smart, it is the complexity of the brain and the relative proportion of forebrain that are most important. The most complex and most forebrain dominated brains on this planet belong to dolphins and humans. Nevertheless we can see that some fish are much smarter than others and the Elephantnose fish is known among aquarists as the most playful and inquisitive of fish.

The brains of cyclostomes (Hagfish and Lampreys) are simple but specifically evolved to suit their lifestyles. For instance the optic lobe is well developed in the visually oriented Lampreys but indiscernible in the blind Hagfish. In both however the medulla is large and the cerebellum small. Together the cerebellum and the medulla make up the hind brain. The medulla controls the operations of the inner organs such as heart rate, blood pressure, digestion and waste disposal. It is also a relay centre for many nerves sending messages to and from the mid and or forebrain. The cerebellum controls motor co-ordination (but it does not initiate motor activities). This means it controls the timing and interaction of muscles once a muscular action has been initiated. The cerebellum is also important in maintaining equilibrium.

The mid-brain of a fish consists mostly of the optic lobes, which vary greatly in size between species in accordance with their dependance on sight, and in some species the optic lobes may be so large they completely cover the forebrain. In fish the mid-brain is important in sorting out incoming information and it is also the main centre of learning (whereas in mammals it is the forebrain that is the main centre of learning).

The forebrain of fish is dominated by the olfactory lobes which extend forwards and may be placed at the end of stalks. These olfactory lobes are large in the cyclostomes and very large in the elasmobranchs reflecting the importance of smell to these to groups of fish. The teleosts, for whom sight is often the most important sense have smaller olfactory lobes. In many elasmobranchs and some teleosts there exists a cerebrum or pair of cerebral hemispheres These also seem to be predominantly involved with the sense of smell (in mammals the cerebrum is much larger and involved in planning and learning). The pituitary also arises out of the forebrain, it plays and important role in the regulation of metabolism.

A fish's brain never completely fills the cranium, the cavity in the skull where it lies protected. The remaining space is filled up with a gelatinous material. Finally as in all vertebrates the brain, plus the gel, are surrounded by a membrane that helps keep foreign matter and micro-organisms from contacting this most important organ.

The Spinal Cord: The spinal cord, or nerve cord is similar in all fish. It is a thick sheath of nervous material that runs from the base of the brain back along the fish's body through, and protected by, the neural canal of the spinal column. Normally it extends the full length of the fish's body, but a notable exception to this is the giant Sunfish (Mola mola) wherein the spinal cord is actually shorter than the brain. It serves as the basis of many simple responses and as the major link to the brain for sensory input and brain-mediated responses.

The Nerves: Apart from the brain and the spinal cord the fish body is supplied with a vast network of nerves, the electric wires of the body along which messages travel. Nerves are built of of numerous neurons and neurons are a one-way system, messages either travel to or from the brain or the spinal cord along a particular neuronal path, but never both ways. Those nerves that arise from the spinal cord are called spinal nerves and those which arise from the brain are called cranial nerves.

The Structure of the Testes and Sperm and the Development Stage of the Testes

The Structure of the Testes

The testes are paired and tubular. They are situated on both sides of the air bladder, attached to the ceolomic wall. Mature testes are white in colour. There are a lot of ampullae arranged irregularly inside. The spaces between ampullae are full of connective tissues.

The ampullae are composed of many spores or seminal vesicle sacs (spermatangium spermatogonia). Spore sacs are separated by a thin layer of follicular cells. In one spore sac there is plenty of synchronously-developing germ cells, and in different spores sacs, germ cells vary in developmental stages. The centre of the ampullae forms a hollow cavity. After the formation of sperms, spore sacs dissolve and the sperms enter the cavity of the ampullae. The terminal end of the testis is connected to a short seminal duct with an opening to the exterior of the body.

The Structure of Sperm

A sperm cell of Chinese carps consists of a head, a neck and a tail. The sperm head of Silver carp is almost spherical, 2.2-2.5 um in diameter, consisting of an apex and a nucleus. The apex is situated at the front part of the head.

It is also called penetrator for its function of penetrating into egg. The neck is very short and is situated between the head and the tail. The sperm neck of Silver carp is about 1.1 um in length. The tail, narrow and long, is many times more than the head in length. The sperm tail of Silver carp is about 35um in length. The tail is the metabolic centre and motor organ.

Mature sperms congregate in ampullae cavities because of the disintegration of spore sacs. Sperms will mix with the fluid secreted by the interstitial cells in sperms and the testis, forming so-called milt. The milt may be exuded or pressed out of body at the climax of brooders' estrus. 1 ml of Silver carp's milt holds approximately 48 million sperms. The total amount of milt could reach 30-40 ml.

The Development Stages of the Testes

Just as the ovary, the development of the testes may be divided into six stages according to its appearance and tissue slice observation.

- *Stage I testes:* Testes are lineal in shape, transparent, attached closely to ceolomic wall. It is impossible to distinguish sexes. On the tissue slice, scattered spermatogonia, 16um in diameter may be observed. The nucleus is big and round, 9 um in diameter. Ampulli and seminal vesicles are still under formation; therefore, there is no clear fixed arrangement of sperm cells.
- *Stage II testes:* Testes are lace-like, transparent or non-transparent. Blood vessels are not clear. Histologic characteristics of this stage are multiplication of spermatogonia, formation of seminal vesicles which are arrayed in bundles. At this stage, ampulli are solid and separated by connective tissues.
- *Stage III testes:* Testes are rod-shaped, pink or yellowish, elastic on surface with clear distribution of blood vessels. On the tissue slice, a hollow cavity may appear in the middle of solid ampullae with one or several layers of seminal vesicles on ampullar walls.
- *Stage IV testes:* Testes are milky white in colour with a clear distribution of blood vessels on the surface. It is impossible to squeeze out milt at early stage, but possible at late stage. On the tissue slice, some larger primary spermatocytes, some smaller secondary spermatocytes and the smallest spermation can be

observed. All these above-mentioned cells congregate on the walls of seminal vesicles with a small number of sperms.

- *Stage V testes:* Testes are white in colour and full of milt. The milt will flow out through cloacal opening if a male fish's head is taken up and its belly slightly pressed. A large number of sperms and sperm cells in different stages could be seen inside ampulli on tissue slice.
- *Stage VI testes:* The capacity of the testis has greatly decreased after milt exudation, yellowish-white or pink in colour on tissue slice. Only spermatogo nia, a little primary spermatocytes and connective tissues remain in seminal vesicles. After exudation, the testies of the male fish is turning back to stage III and will develop on.

Sexual Cycle and Maturity age of Fish

There is a certain sexual cycle in the development and maturity of the gonads of fish. When fish reach the maturity age and have their first spawning or milt discharging, their gonads will develop cyclically on without ceas in the light of season changes which is termed sexual cycle. In pond culture, the sexual cycle of Silver carp, Bighead, Grass carp, Black carp and Mud carp are basically the same. But, in nature, they spawn once a year, so the sexual cycle is one year. In the southern part of China, the climate is rather warm. Through intensive culture, pond-reared spawners after spawning in spring can again reach maturity in the same year and be induced to spawn 2-3 times a year.

The sexual maturity and the gonad development of fish are tightly associated with environmental conditions that is water temperature, feeds, oxygen etc. Accordingly, there is a variance of fish sexual cycle in different regions. Even in the same region, the gonad development could not be totally the same owing to various conditions.

Cyclic Variance of Ovary Development

In winter (November-next January), the ovary is at stages I-III with average maturity coefficient 5-6 per cent. Huge amount of lipid in fish body occupies about 3.5 per cent of the body weight. Eggs of old generation are at the stage of degeneration and absorption, some of them have already been absorbed. The new oocytes begin to accumulate egg yolk.

In summer (May-July), the ovary grows fast with obvious increase of weight, having passed through stage IV—maturity stage with maturity coefficient 17-20 per cent. The lipid content of fish body reduce sharply; the belly is soft and expanded. If water temperature is favourable, inducement may be conducted. In autumn (August-October), already after ovulation, the ovary is in atrophy with the maturity coefficient around 10 per cent. The majority of those ovaries that have not ovulated show apparent degeneration, returning quickly from stage IV to stage II. In winter, ovaries will develop again from stage II or stage III. The nutritive materials such as lipid will be gradually accumulated for the next sexual cycle. As Mud carp are subtropical fish, the gonad development requires higher temperature. The gonad develops slowly in winter when ovary is at stage II. Upto next April (water temperature 26°C in the southern part of China), the gonad develops quickly into stage IV.

Cyclic Variance of Testis Development

The development of male fish's testis is earlier than that of female's ovary. Generally speaking, it has already reached stage III in water and could exude a little milt under pressure in early spring and before reproduction season. In autumn, after milt-discharging, the sperms remaining in testis will be absorbed through degeneration. The spermatocytes of a new generation begin to grow. The testis takes another new sexual cycle.

Maturity Age

Under different geographical and ecological conditions, the maturity age of the same species is widely different. The maturity age of Silver carp, Bighead and Grass carp in the southern part of China is 1-2 years earlier than that in the northern part of China both for the male and female.. And yet, even in the same region, it differs more or less under variant ecological conditions, for instance, in Jiangsu and Zhejiang Provinces, the maturing age for female Silver carp is 4 years, for Grass carp and Bighead 5 years, Black carp 7 years. If ecological conditions are suitable, the maturing age could be one year in advance. The males generally mature 1 year earlier than the females.

Fecundity

It refers to the number of mature eggs of a female fish before spawning or individual fecundity. The assessment of the fecundity of fish is based on maturing age, sexual cycle, brood amount, effective production of eggs and the survival rate of fry, etc. Emphasis here is laid on the brood amount and spawning amount of pondreared Silver carp, Bighead, Grass carp and Mud carp. Production of eggs: Under artificial propagation, the average egg production of Silver carp, Bighead, Grass carp and Black carp is 50 pieces per gram of body weight. The highest is 100 pieces/g. The average amount of Mud carp is a bit higher because the eggs of Mud carp are much smaller..

Hearing and Sound Production in Fish

Sound travels further and faster in water than it does in the air, about 4.4 times as fast. Because of this, and because visibility is greatly reduced in water, it is not surprising that sound is an important part of how fish communicate with each other and with the world around them.

The seas, lakes and rivers of the world are often alive with a riot of noise, but not all of this is made by fish. Besides other organisms there is also the noise that water makes as it moves against solid objects, or is moved by the wind and in our modern world the noise made by mankind and his machinery.

Many things can effect the speed of sound, including not only the nature of the medium, (gas, liquid or solid) but also its temperature and any other additive substances, such as salt in water. Basically sound travels faster through denser and hotter materials.

Medium	Temperature	Speed in M/s
Air	0	331.4
Air	20	343.6
Air	30	348.7
Fresh Water	Unknown	1,493
Sea Water	Unknown	1,533
Diamond	Unknown	12,000

Like us fish produce sound in two main ways, intentionally and unintentionally. Unintentional sounds are produced by fish all the time, mostly by swimming and feeding. However they make a far greater variety of sound intentionally in their efforts to communicate with the other creatures living in their world.

The sounds that fish make are usually simpler than the complex songs of birds, or the calls of mammals. Scientists usually describe them as grunts, scrapes, knocks, clicks, squeaks, groans, booms, thumps, rumbles and drumming. Nearly all the species studied so far produce their sounds using either their teeth, their swimbladder or a combination of both. Those species that produce sounds using their swimbladder usually have special muscles attached to it for exactly this purpose.

Fish create sounds for several different reasons, to stay in touch with the shoal, to warn shoal-mates of danger, to attract, communicate with and stimulate mates, to scare intruders away from eggs and young and possibly even to echolocate in some deep-sea species.

Some fish are capable of making very loud sounds. One of the noisiest fish in the oceans is the Oyster Toadfish, Opsanus tau. Because of their noisiness Oyster Toadfishes were studied by the US Navy, they kept hearing them on their sonar, and it has been claimed that measured from a distance 60 cm the volume of sounds produced by the Oyster Toadfish can reach 100 decibels, which is equivalent to a piece of heavy machinery.

Of course many fish try to take advantage of the sounds other species make. Thus some sharks will use sound to help them locate their prey while some smaller fish can detect the sounds larger predators make in their hunting. Recent research has shown that some Clupeid fish, (Herrings and Shads) can detect the ultrasonic echolocation sound produced by hunting dolphins from a distance of up to 187 metres.

Hearing

Fish hear with their ears, which, although similar to ours in their basic functionality are missing both the pinna (earlobe), and the middle ear with its eustachian tube. The inner ear, which is responsible for both hearing and balance is located behind the eye. The modern fish's ear consists of 3 chambers called the 'utriculus', the 'sacculus', and the 'lagena'.

The utriculus, and the 3 semi-circular canals associated with it, are entirely concerned with balance. The three semi-circular canals are arranged in three different planes of orientation, just as in the mammalian ear, one horizontal and two vertical, but at right-angles to each other. While this is true for the Osteichthyes and the Chondrichthyes the more ancient Lampreys have only two canals, the horizontal one is absent, and the hagfish have only a single canal.

These canals, which are sometimes collectively called the Labyrinth, have a bulbous area at each end called the ampulla. These ampullae each contain the cupula. Movement of the fish's body, in any plane, causes movement of the fluid inside the canals. The movement of the fluid causes the cupulae to move, when the cupulae move they press against sensitive hairs, which in turn send electric impulses to the animal's brain.

The 'utriculus', the 'sacculus', and the 'lagena' each contain an 'otolith' or ear-stone. This is normally calcareous and therefore much denser than the lymph fluid that fills the chambers or even the fish's flesh. Movement in the lymph fluid within the chambers, caused by sound waves, causes the otolith to impact against sets of sensitive hairs within the chambers, which then also send messages to the fish's brain.

Expressed again in slightly different way, the otoliths hang in their chamber of fluid, passing sound waves cause the molecules of the water to oscillate, the exact movement is dependant on amplitude and wavelength of the sound waves, this movement of the water molecules causes the fish to rock back and forth in the water a little and because the otoliths are denser than the lymph fluid they move less and or more slowly and thus impact on the sensitive hairs. It is interesting to note that the shape of the otoliths is different for nearly every species of fish, and scientists can often identify a species of fish just from the otoliths, quite why this should be so is a mystery waiting for a scientist to unravel it.

The basic fish ear is fairly limited in its range of use. However many teleost fish have their inner ear connected to their swim bladder, either directly, as in fish like herrings, or indirectly via a chain of small bones called Webarian Ossicles. This allows them to use their swim baldder as a sort of drum to detect a greater range of sounds.

In the cartilaginous fish, which have no swim bladder, this is obviously not possible. Instead they have duct, filled with endolymph, that connects the sacculus to the external environment. This duct passes through an area called the parietal fossa, this parietal fossa is covered by a layer of skin which may act as a sort of ear drum.

Nevertheless Sharks and Rays have a relatively low hearing range, even though they have been shown to directionally locate sources of low frequency sound pulses from up to 100 ms. The shark's ear also contains an organ called the Macula Neglecta whose exact function is not known, however it is suspected that it may well respond to movements in lymph fluid caused by sound waves contacting the skin that covers the parietal fossae.

Hearing Ranges in Fish

Fish can hear a wide range of sounds, however there is a great variety of capabilities between species. The data as shown in table has been collected from a number of papers and serves well to demonstrate great variability in hearing potential within the world of fish.

3

Biotechnology in Fisheries Development

The gap between human demand and the availability of livestock products is huge in Africa, and imports have been increasing fast. The productivity of African breeds is very low: mature beef cattle of 4-5 years hardly weight 300 kg (as against 400 kg at one year for exotic animals); the best African cows produced 300 litre/lactation (as against 5 000 litres and more in Europe); 100 African ewes produced 50 lambs every two or three years (as against 150 lambs a year in developed countries).

Retrospectively, production policies that aimed at developing livestock through cross-breeding of exotic grade cattle with indigenous ones have generally failed.

The new approach is not to make selection of local performant animals. Biotechnology offers good methods both for selection and cross-breeding schemes, essentially through Multiovulation Embryo Transfer (MOET) and the field diagnosis of pregnancy. Yet, there is wide diversity among the countries to use new technologies.

For instance, even if simpler techniques such as Artificial Insemination (AI) are considered, it is seen that in East Southern Africa the technique is widely used, whereas in West and Central Africa, because of the predominance of nomadic transhumant livestock production, the technique is least developed.

MOET is well developed in few countries such as Zimbabwe and South Africa. Cattle farmers in these countries import frozen embryos from overseas. MOET is promising especially in the development of dairy cattle farms around large cities. Both AI and MOET are being used in Open Nucleus Breeding Systems (ONBS) to improve animal production. Small-scale ONBS projects are being carried out by FAO in small ruminants in Ghana and the Gambia.

Moreover, by using semen from bulls, with genetic resistance to high temperature, diseases and insects common to African countries, embryos could be produced from superior cows of other regions and then implanted into surrogate mothers within the environment where improved livestock are needed. In animal feeding, biodegradation of low-quality forage is also a promising technology. It can be associated with chemical degradation techniques (use of urea and ammonia) to improve animal feeds during dry seasons.

In some of the countries, farm diagnosis of pregnancy is also being used by field workers by using a kit based on the difference of levels of progesterone in cow's milk.

In the field of animal health, Sterile Insect Technology (SIT) is well known in Africa. Sterile males of glossina are being reared in Bobo-Dioulasso and Burkina Faso for field application. Successful field projects in tsetse control have been carried out in Nigeria (BICOT Project) and in Burkina Faso. Newworld screwworm was eradicated in the Libyan Arab Jamahiriya in 199091 by the use of SIT.

In disease diagnosis, ELISA is being used to evaluate rinderpest immunity in vaccinated animals. Monoclonal Antibody Techniques (MAB) is not commonly used. The two techniques are very promising because they are very simple and most effective. Trials are under way to use the ELISA test in trypanosomiasis diagnosis. In vaccine production recombinant rinderpest vaccine trials done in the United States seem ready to be utilized in a mass vaccination campaign against rinderpest in Africa. There is an increasing level of activity in hybridoma work in sub-Saharan countries.

The work of ILRAD in Kenya is addressing the problem of diseases and parasites that affect cattle. Some of the common cattle diseases are East Coast Fever, tickborne diseases and trypanosomiasis. The focused approaches to researching on these diseases is expanding to yield results that will significantly benefit cattle breeders. The development of rinderpest vaccines remaining stable at ordinary temperatures was of high practical value.

Biotechnology in fisheries is recent in Africa and it is being applied essentially in private fisheries sector. Work is mostly confined to hormonal treatments for sex reversal and pawning, detection of fish and shrimps disease and development of fish feed.

Developments in Biotechnology

There exists a gradient of biotechnologies, depending on the degree of sophistication, complexity, stage of development and application. The lower side of the gradient comprises simpler but widely used techniques such as in vitro culture, rhizobium technology, fermentation. The upper side of the gradient includes advanced techniques involving genetic engineering. The gradient of biotechnologies may be matched with the gradient of national capabilities, economic investment and efforts to provide the possibility of choosing the appropriate techniques and approaches with the most positive impacts. The pattern of use and development of biotechnologies in developed and developing countries thus varies considerably.

The Pattern in Developed Countries

In the industrialized countries, a new pattern of biotechnological research funding has emerged. With the availability of property right protection of biotechnologies and prospects of vast markets for biotechnology products and the techniques, the bulk of the research is funded, carried out, and controlled by the private sector. For instance, since 1976 when GENENTECH, the first biotechnology company, was established, the new techniques have spawned a variety of industries that now comprise more than 400 start-up firms, more than 200 established firms that have diversified into biotechnology, and more than 200 supply firms in the United States alone.

The new biotechnology industry in the United States produced pharmaceutical, diagnostic tests and agricultural products worth close to US$2 billion in 1990. A similar trend is seen in Europe and Japan. It is estimated that about 60 percent of the biotechnology research and development funding in industrialized countries is from the private sector. Thus, the private sector has been a major force in enhancing the capability of these countries in this field.

Research institutions in the public sector are now generally required to raise a substantial part of their budget from non-government sources, via contractual research, licensing agreements and royalties. This is tending to increase secrecy over research findings and to hinder free scientific communication. University professors, researchers and government institution scientists are increasingly becoming entrepreneurial and are entering private industry.

Another important trend is that large multinational corporations are purchasing smaller seed and biotechnology companies and diversifying their holdings. This allows them to develop a package sale of chemicals, seeds and equipment.

Heavy involvement of the private sector and market considerations greatly influence the topics and commodities chosen for research. Major crops, commodities and farming systems of great socio-economic importance to the developing world, but of little international market importance, do not figure in the biotechnology research agenda of industrialized countries. Furthermore, these countries are keen to reduce their production costs, increase the productivity, quality and value of their products and, thus, improve their overall competitiveness in the world market.

The Patterns in Developing Countries

Biotechnology facilities are being established in most developing countries. However, the level of research, development and use of biotechnology for agriculture, forestry and fisheries in the developing countries is generally far below the level in the industrialized countries. Among developing countries, the status varies considerably. A few, such as Brazil, Mexico, India, China and The Republic of Korea, have sought to gain full scientific and technological capacity, especially in agricultural biotechnology. Others, such as Indonesia, Malaysia, the Philippines and Thailand and a few countries of Latin America, have built the capacity to apply biotechniques and develop biotechnologies useful for their agriculture and food industries. The participation of the private sector in gaining biotechnological capacity is not significant in most of these countries.

Many developing countries have inadequate funding, poor human resources and limited access to information, resulting in a relatively low-level capacity for research and technology development and exploitation, especially in the field of modern biotechnology research, which is costly and requires highly trained personnel. Most developing countries are vague as to their immediate aims in agricultural biotechnology. Few have appropriate proprietary-right protection systems or mechanisms to increase their access to protected techniques and products. Furthermore, there is negligible involvement of the private sector, which accentuates the problem of insufficient attention to biotechnology.

One of the major constraints on biotechnological development in developing countries is the poor quality and extent of higher education in frontier sciences, especially molecular biology.

Besides, there are no or very weak links between universities and research institutions to reinforce each other and to utilize synergistically the scarce trained personnel resources. The university experts play a negligible role in national policy formulation, including that on biotechnology. Furthermore the universities are generally not oriented for efficiency in commercialization and are therefore not able to market biotechnology products of their own or others' inventions.

Some international and donor-country public sector institutions strive to ensure that attention is given to solving, through biotechnology, the food and agriculture problems of developing countries. Among this group are the International Agricultural Research Centres (IARCs) located or operating in developing countries. Capacity exists in several of these centres for utilization of advances in biotechnologies to solve certain problems of plant production and protection and of livestock production and health, and capacity is being built in others.

For example, the International Rice Research Institute (IRRI) is incorporating modern molecular biology in its rice research activities. Its programme is linked closely to the Rockefeller Foundation Biotechnology

Network. The World Bank, the International Service for National Agricultural Research (ISNAR) and the Australian Government undertook a major study on the likely impact of modern biotechnology on agriculture. This study is being followed up by the participants by an expansion in the World Bank lending for biotechnology and by ISNAR's establishing an Intermediary Biotechnology Service to provide advisory services to national agricultural research systems.

Applications, Impact and Potential

Biotechnological research and development are moving at a very fast rate. For instance, three years ago, transgenics in rice, especially in japonica rices, were considered rather difficult because of the problem of regeneration through protoplast culture. But today, a large number of transgenics both in indica and japonica types are available and are being tested at various stages. It is therefore difficult to predict the potential and impact of a given technology beyond five years or so. This section gives a brief account of the current level of technology and refers to the developments likely to occur in the next three to five years.

The potential of genetic engineering in medicine has received much attention. The potential for advances in agriculture, forestry and fishery are similarly bright. Biotechniques are already being used to create new strains of crop plants, new plant and animal diagnostic products, animal vaccines, biological pesticides and herbicides, other biological control agents, and modifications in domestic animals used for food production.

In some cases, applications are being held in check by the need for still more research to ensure that there are no harmful effects (or, if there are, they are greatly outweighed by the benefits) and by the slow pace of evolution of regulations governing the release of genetically modified organisms or products for general use or acquisition.

Biotechnological research, especially genetic engineering, on problems of field and tree crops, forest species and fisheries has a relatively short history. The knowledge of biological aspects of species and ecosystems needs to be expanded in may cases before new biotechnologies can be applied effectively. Even where transfer of one or more genes can be achieved reliably, expression of the transferred gene(s) may not occur in the expected way. The search for and identification and cloning of useful genes, effective vector systems, methods of gene transfer and promoter mechanisms should be intensified.

A perspective on both the difficulties and intensity of the research effort suggests that diagnostic tools will speed solutions in the next five to ten years to certain disease problems in several crop, forest, animal and fish species, such as black sigatoka of banana and plantain, and virus/viroid diseases of coconut and rice. Improved tolerance to certain physical stresses, such as heat tolerance in potatoes, may also be achieved, but prospects for improved resistance of varieties and clones to most abiotic stresses are longer term.

Livestock

Among agricultural and allied fields, animal production and health have benefited the most from biotechnology, although practical use of transgenic livestock is only a future possibility. Wide use of monoclonal antibodies for efficient diagnostics, leading to safe and specific treatments of animal diseases, is a major breakthrough.

Through genetic engineering, vaccines for the prevention of viral, bacterial, and parasitic animal diseases are rendered more effective and safe. Tailored vaccines exist for pig scours, chicken bursar disease and cattle tickborne diseases. Pathogen-specific vaccines are attractive goals. Other interesting possibilities are endocrine-directed vaccines to stimulate twinning in beef cattle, immunocastration, livestock growth rate increase and vaccines that compensate for various stress-induced production losses.

Advances in genetic engineering will also facilitate the production of male-only populations of screwworms, tsetse flies, ticks and various other ectoparasites for use in the sterile release technique of control or eradication. Furthermore mammalian tissue culture may replace whole animals in the 1990s for toxicity testing of certain chemicals. The culture technique can also be exploited for studying and analyzing pesticide metabolism and for herbicide pre-screening. In vitro fertilization and embryo sexing techniques have considerably increased the use of embryo transfer techniques for cattle breeding and trade. The value of the approach will be further enhanced if embryo cloning techniques can be reliably employed. Microbial and enzymatic treatment of roughages and genetic engineering of rumen bacteria both have great potential to improve animal nutrition. Growth hormones can be produced by genetically engineered microorganisms in quantities and at the low cost necessary for widespread use to speed and increase milk and lean meat production. Biotechnological tools (embryo culture, gene cloning, etc.) may also be used for conservation of genetic resources.

Forestry

Successful regeneration through micropropagation or somatic embryogenesis has been achieved for about 100 forest species although, for most, considerable development work would be required before commercial propagation could be contemplated. In breeding programmes incorporating clonal testing, inclusion of a micromultiplication phase may facilitate more rapid deployment of superior genotypes than that afforded by sequential multiplication by cuttings. This will particularly be the case when gene mappling techniques are sufficiently advanced to permit accurate identification of superior genotypes without replicated clonal testing in the field, especially within a breeding line where substantial genetic disequilibrium exists. Furthermore, gene mapping may allow eventual identification of valuable genes, including those that contribute to quantitative traits.

As regards the application of genetic engineering, many major traits of commercial importance to forestry are under polygenic control, and much remains to be learned about the operation of the genes involved before this technique can have a major impact for these characters. An earlier application of genetic engineering in some forestry operations may be the introduction of genes, some already available, known to confer insect of disease resistance. Other applications of biotechnology with obvious value for forestry, but as yet not widely supported by experimental successes, include: in vitro manipulation of the maturation state, *e.g.* the promotion of early flowering to reduce generation intervals, in vitro selection for traits such as disease resistance and tolerance to salinity; and the use of haploid cultures.

Fisheries

Two broad areas of biotechnology exist within the fisheries sector, namely natural products biotechnology (including mostly marine species); and aquaculture biotechnology. Commercially valuable products such as pharmaceutical compounds, pigments, oils, sterols, alginates and agarose are being extracted from micro-and macro-algae in many parts of the world. Aquatic invertebrates are currently being screened for biologically active compounds that may have, inter alia anti-carcinogenic or antiviral properties, whereas marine bacteria are currently utilized in treatments of oil spills and holding tanks in tanker ships.

Within aquaculture, induced increases in the chromosome complement (polyploidy) of commercially important species such as salmon and oysters have increased the growth and marketability of these species. Genetically engineered microbes can be used to produce fish growth hormones, which might then be used to improve feed conversion and growth rate. Synthetic reproductive hormones are produced commercially and used to regulate fecundity, breeding cycles, growth rates and sex determination in certain cultured species.

In attempts to increase desirable culture qualities such as growth rate, disease resistance, temperature tolerance and marketability, transgenic fishes containing introduced genes from other species have been produced on an experimental scale. The application of advanced biotechnology to the fisheries sector is a relatively recent practice. Therefore, continued research can be expected to reveal additional commercially viable applications and products within this sector.

Development of Fish Embryology

Fish embryonic development begins after inception with the zygote period. During this time the embryo develops one cell that looks like a half bubble. This lasts on average for 45 minutes with a cell .7 mm in diameter. The chorion, also known as the egg shell, will swell and lift away from the fertilized egg.

It is particularly appropriate to write about the history of fish embryology at this time. History is often said to move in cycles, and this seems to be true of the study of fish development.

After a period of prominence during the nineteenth and early twentieth centuries, the study of fish development, except in the hands of a stalwart few such as Oppenheimer and Trinkaus working with Fundulus and Ballard and Devillers working with the trout and other fishes, was overshadowed by developmental studies of other organisms, such as amphibians, the domestic fowl, sea urchins, and assorted invertebrates. For obvious reasons, the study of fish development continued to be pursued by fisheries biologists. With the popularization of the zebrafish Danio rerio as a "model of vertebrate development" and a revival of interest in the relationship between development and evolution, the cycle has come full turn and the study of fish development again has a place in the sun.

My objective is to present an outline of the historical development of the science of fish embryology in the nineteenth century, broadly defined as 1789-1914. My plan is to assess the state of knowledge at the outset of the period, to document the advancement of descriptive embryology of teleosts and elasmobranchs, to consider the rise of comparative embryology of fishes associated with evolutionary studies, and to search out the beginning of experimental, physiological, and biochemical analyses of fish development.

Diverse group of vertebrates

When discussing the development of fishes, it is prudent to recall that fishes are the most diverse group of vertebrates. They should not be treated as a distinct, relatively homogeneous taxonomic unit. Rather, the term "fishes" applies to a grade of ectothermic, aquatic craniates and vertebrates and not to a distinct taxonomic group. Extant fishes are divided into five classes, namely Myxini (hagfish), Cephalaspidomorphi (lampreys), Chondrichthyes (cartilaginous fishes), Sarcopterygii, and Actinopteryg. Chondrichthyes contains two taxa, namely the subclass Elasmobranch (living sharks, rays, and skates) and the Holocephal (ratfishes or chimaeras) that diverged at a relatively early time, *i.e.*, the Devonian-Carboniferous boundary.

The class Sarcopterygii (flesh-finned fishes) contains the Actinistia (the living coelacanth), the Dipnoi (lungfishes) and the tetrapod vertebrates. The class Actinopteryg (ray-finned fishes) comprises five groups, namely Cladistia (bichir and reedfish), Chondrostei (sturgeons and paddlefish), Ginglymodi (gars), Halecomorphi (Amia calva, the bowfin), and the Teleostei (teleost fishes). Thus, the five classes of fishes contain eleven major taxonomic groups with an estimated total of 25,000 species. Each class is as distinct from the others as it is from the four other classes of vertebrates. The theme of eleven major taxonomic groups of fishes that is used in the ensuing presentation to emphasize the broad phylogenetic relationships of fishes and other vertebrates, tends, however, to obscure the extraordinary phylogenetic diversity within the teleosts.

Teleostei, a monophyletic group, contains 24,000 species in 38 orders and 126 families compared to the 1,000 species, 19 orders, and 56 families in the ten other groups of fishes and the 25,000 species of tetrapods. Not only are fishes the most diverse group of craniates, but they also display the greatest diversity of life history styles. Most fishes are oviparous, but viviparity is estimated to have independently evolved 42 times in five of the eleven major groups of fishes, all within the Chondrichthyes and Osteichythes. The phylogenetic diversity of fishes is reflected in the diversity and variability of their developmental patterns. Craniate development based on cleavage and subsequent patterns of gastrulation and embryogenesis is holoblastic, meroblastic, or transitional between the two. In holoblastic eggs, cleavage of the ooplasm is complete by the blastula stage. The morphogenetic movements of gastrulation occur in the three-dimensional space of the entire egg and involve most of the embryonic mass.

They often lead to the occlusion of the blastocoele. Neurulation is synchronous in time and space. Holoblastic development is considered the less specialized and more primitive craniate developmental pattern because it is characteristic of most echinoderms and non-craniate chordates as well as the more primitive representatives of various craniate lineages, namely lampreys, sturgeons, bichirs, lungfish, and most amphibians. In the meroblastic egg, most of the yolk mass remains uncleaved at the blastula stage.

Evolution of the meroblastic pattern appears

Morphogenetic movements are confined to a portion of the egg volume, *i.e.*, the blastoderm, and involve only a portion of the embryonic mass. Gastrulation and neurulation tend to be spatially and temporally asynchronous. Evolution of the meroblastic pattern appears to have involved qualitative and quantitative changes in the ooplasm and yolk components of the egg, namely an increase in total yolk content, an increase in the amount of yolk relative to ooplasm, a change in the type of yolk, and a change in the relative disposition of ooplasm and yolk. Not only did the changes in yolk-cytoplasm relationships of the egg affect cleavage patterns, but they also imposed a set of physical or physiological constraints on the morphogenetic movements of embryo formation, the so-called holoblastic-meroblastic barrier.

Consequently, new or highly modified patterns of gastrulation and embryogenesis co-evolved with the evolution of the meroblastic egg. The meroblastic pattern of development appears to have independently evolved in six craniate lineages, namely hagfish, sharks and chimaeras, teleost fishes, the living coelacanth (?), one group of amphibians (the caecilians) and amniotes. In each instance, different morphological patterns of gastrulation and embryogenesis have evolved. Thus, although meroblastic cleavage is superficially similar in hagfish, elasmobranchs, and teleosts, the morphological events of embryo formation differ considerably. These points and others are summarized in part by Collazo *et al.* who present a phylogenetic perspective and an evolutionary scenario for teleost gastrulation. Paradoxically, whether development proceeds according to a holoblastic or meroblastic pattern, the end product is a stereotypic craniate embryo. Recent advances in the study of the molecular basis of embryogenesis should help resolve this paradox.

Natural collections of fishes

When considering early studies of fish development, one soon realizes that many of these studies were carried out on viviparous species of elasmobranchs and teleosts rather than oviparous ones, and that both egglaying and livebearing elasmobranchs received a disproportionate amount of attention, considering their relatively low diversity compared to teleosts. First, elasmobranch eggs are considerably larger than those of most teleosts and their development can be easily observed with either the unaided eye or with a hand lens. Secondly, it has always been relatively easy to obtain gravid specimens of viviparous fishes as part of routine fishing efforts. Moreover, as museum or natural history collections of fishes were made during the course of explorations, specimens of curious, new species of gravid viviparous fishes became available for study, e.g., the four-eyed fish Anableps. In contrast, the eggs of oviparous teleosts are small and not easily collected in the wild. Artificial fertilization of teleost eggs did not begin until the late eighteenth century and the regulated spawning of captive fish species and the rearing of their eggs under "controlled" conditions for scientific purposes was first used only towards the mid-nineteenth century.

In addition, the technology for maintaining fishes in aquaria and the possibility of spawning them under aquarium conditions did not emerge until the mid-late nineteenth century. In passing, one should recall that the Chinese have been rearing captive goldfish, including mutant strains, for over a thousand years. Other factors also influenced decisions to study certain viviparous instead of oviparous fishes. Viviparity in fishes, especially when there is a placental relationship, is an intrinsically interesting process. Finally, as a subject for the general study of development, oviparous teleosts have been secondary to the chicken egg ever since the Renaissance because of the greater size of the chicken's egg and its ready availability as an article of food.

Development of fish embryology

Because the research of early naturalists provided the foundation for the development of fish embryology in the nineteenth century, a brief outline of their work is presented here, beginning with Aristotle. Aristotle, the founder of the science of embryology, was the first to recognize and record many important features of fish reproduction and development. He distinguished oviparous (egg-laying) animals from viviparous (livebearing animals. Aristotle described viviparity in the dogfish shark (Mustelus), the thresher shark (Alopias), and the electric ray (Torpedo). He did not know of viviparity in teleost fishes because there are no viviparous teleosts in the coastal waters of the Mediterranean or the fresh waters of Europe and Asia Minor. His attribution of viviparity to the needlefish, a teleost, is actually a case of misinterpretation. The needlefish of Aristotle is a pipefish or sygnathid.

Male pipefishes brood embryos in a pouch of abdominal skin. This form of parental care is analogous to viviparity and in its most highly evolved forms, the parentalembryonic relationships can be as complex as in viviparous fishes. The body of Aristotle's observations concern chondrichthyan fishes and include: (1) the distinction between oviparous and viviparous modes of reproduction in sharks, rays, and skates; (2) description of the female reproductive system; (3) description of shark and skate egg cases; (4) description of egg structure and some observations on embryonic development; (5) first description of the oviducal or nidamental (=shell) gland but without any knowledge of its function in egg case formation or sperm storage; (6) hatching from the egg case in utero in viviparous species; (7) description of the male reproductive system; (8) notes on breeding seasons and inshore migration of gravid females for "pupping"; and (9) description of the yolk sac placenta in sharks. Writing of the smooth dogfish Mustelus canis in the 5th century B.C., Aristotle states, "the young are produced with the umbilical cord attached to the uterus so that as the substance of the egg gets used up, the embryo's condition appears to be similar to what is found in quadrupeds. The umbilical cord, which is long, is attached to the lower parts of the uterus; each one is, as it were, fastened to a cotyledon and is attached to the embryo by the middle where the liver is situated. Each embryo has a chorion and membranes of its own round it, just as in quadrupeds. If the embryos are cut open, a situation is disclosed exactly similar to quadrupeds; whatever internal organs they have, such as the liver, are large and supplied with blood".

Aristotle's observations on teleosts are in large part concerned with reproduction, e.g., his exquisite account of brood protection by male catfish. He realized that in contrast to the chondrichthyans, the overwhelming majority of teleosts are oviparous. His observations on the development of teleosts are more limited, due primarily to the small size of most teleostean eggs. Aristotle observed developing teleost eggs, recording their small size and rapid development. He states, "The embryo appears at the upper end of the egg and is enveloped in a membrane; the eyes, large and spherical, are the first organs visible. The embryo and the egg (yolk) are enveloped by a common membrane, and just under this is another membrane that envelops the egg (yolk) by itself; between the two membranes is a liquid". He described the organization of the teleost egg, distinguishing the perivitelline fluid from the yolk mass. Lastly, in comparing the development of teleostean and avian eggs, Aristotle recognized that teleosts lack the allantois. Relatively little was added to the pioneering efforts of Aristotle by subsequent Greek and Roman scholars. With the decline of classical civilization, the study of science in general as well as the study of fish development languished for almost 1,000 years.

Slowly the tide turned. According to Needham (1959), the first account of fish development in medieval Europe was given by Albertus Magnus, a thirteenth century scholastic. During the Renaissance, there was a revival of interest in natural history. The early zoological encyclopedists, especially Belon, Rondelet, and Aldrovandus, investigated fishes and described aspects of their reproduction and development. Again, there was a decided bias towards viviparous chondrichthyans. Rondelet illustrated an egg case and dissection of the female reproductive system of an oviparous catshark. He also illustrated part of the ovary and an egg case of a skate. He depicted a female shark, probably Mustelus canis, connected to a welldeveloped pup by an elongated yolk stalk that passes from the pup through the cloaca of the female. Although the placenta is not illustrated, Aristotle's research on the shark yolk sac placenta is cited in the text.

Progressing to the seventeenth century, Fabricius, who is best known for his studies of the chick, described and illustrated the female reproductive system and uterine embryos with large yolk sacs in a non-placental species of dogfish shark. By 1673, Niclaus Steno had rediscovered the yolk sac placenta in the smooth dogfish, M. canis, investigated its anatomy, and published an illustrated account of it. He also distinguished between placental viviparity in the smooth dogfish shark, and aplacental viviparity in the spiny dogfish shark, Squalus. In the latter, he discovered that the yolk stalk terminates in an intestine characterized by a spiral valve.

His rediscovery appears to have gone unappreciated. During this period, inability to distinguish among species of dogfish sharks that had either aplacental or placental modes of viviparity caused the very existence of a yolk sac placenta in sharks to be doubted. This controversy was not fully resolved until the time of Johannes Muller. Steno's studies on shark development had important ramifications for the study of mammalian reproduction. On the basis of his observation of large eggs in the uterine oviduct and large eggs in the ovary of the spiny dogfish, he concluded that the embryo does not form de novo in the oviduct, but comes form an egg released by one of the ovaries. From this, he postulated that mammalian development proceeded in a similar fashion, except that the mammalian egg is much smaller.

In his examination of the mammalian ovary, unfortunately, he mistook the entire follicle for the mammalian egg. These studies were undertaken at a time when a controversy had arisen between animalculists and ovists with respect to the primacy of the sperm in reproduction and the absence of any knowledge of the mammalian egg. Steno's observations, although flawed, brought mammalian development in line with that of other vertebrates. In essence, he used sharks as a "model" system for mammals. In 1678, Stephen Lorenzini, a pupil of Steno and Redi, published a set of observations on the viviparous electric ray Torpedo. This appears to be the first monographic treatment of a single fish species. Lorenzini's work is of considerable interest because almost one third of it is devoted to an account of urogenital anatomy, reproductive biology, and development of Torpedo, as well as several other rays.

Amongst his observations, those of primary interest concern oogenesis and ovulation, the blastoderm, and embryonic nutrition. Describing the ovary, he notes differences in size and structure between young and mature fishes. Maturing oocytes and ova vary in size, composition, and colour. Oocytes are connected to the ovarian matrix by a stalk through which several blood vessels pass, an arrangement similar to that described by

Fabricius in the hen's egg. He states that these vessels provide nutrients to the eggs.

Pole constitutes

The maturing egg is invested by a "membrane" except for the pole opposite the site of attachment to the stalk. This pole constitutes the pore or stigma through which the mature egg is released from the follicle. During oocyte maturation, changes in membrane adhesion were documented. Lorenzini speculates that mature eggs are released from the follicle by contraction of fibres in the ovarian stroma. The ovulated egg is "received" by the oviduct and passes into the uterine region where embryonic development is completed. Following Fabricius, De Graaf, and Harvey, Lorenzini refers to the residual portion of the follicle as the "calyx." Calyces, which apparently are the corpora lutea, occur in various sizes and shapes throughout gestation, and are lost or regress some time after parturition. Lorenzini found a correlation between the number of young in a brood and the number of calyces in the ovary. Ovulation in the Torpedo is considered to conform to De Graaf's account in other species. Among the more interesting of Lorenzini's observations are those made on the early stages of embryonic development. He describes and illustrates two neurula stages.

The earlier stage depicts an early neural groove on the surface of the blastoderm. He describes the characteristic chondrichthyan blastoderm, "a leaden gray disc" that is clearly distinct from and surmounts the "yellow yolk mass." In another egg, he describes and depicts a slightly more advanced neurula stage. Here Lorenzini correctly shows the splayed, outspread cephalic margins of the medullary plate, a characteristic feature of chondrichthyan embryos. Lorenzini also observed the blastoderm in three other species of rays. Lorenzini seems to be the first to have depicted the chondrichthyan blastoderm. His observations appeared only six years after Malphighi's initial account of the avian (chick) blastoderm.

Lorenzini made one of the first major contributions to the knowledge of embryonic nutrition in non-placental viviparous fishes. Late stage embryos with yolk sacs lie free in the uterine region of the oviduct where they are bathed by uterine fluid. A yolk stalk passes from the external yolk through the abdominal wall and empties into a large internal yolk sac that in turn empties into the intestine. The intestine is filled with yellow material apparently derived from the yolk and brown globules. The same brown globules occur in the uterine fluid and in the mouth, esophagus, and stomach of the embryos.

Lorenzini concluded that advanced embryos are nourished both by yolk passed to the intestine and by uterine fluid which they imbibe. He believed that the nutrients in uterine fluid are derived from the blood and conveyed to the uterine fluid by vascularized glandular structures which he compared to chorionic villi. These structures are now known as trophonemata. Villi increase in size during gestation; their size is correlated with the amount of uterine fluid. They regress and are absent in non-gravid fish. In addition, Lorenzini also made significant observations on the male reproductive system, reproductive cycles, and the duration of gestation.

In the closing years of the seventeenth century, Willughby and Ray summarized much of what was known of fish reproduction and development in a work primarily devoted to systematics. Considering its promising beginning, it is surprising that the study of chondrichthyan development made relatively little progress in the eighteenth century. There are many possible reasons. Those scholars interested in fish biology may have been attracted to systematic studies because of the revolutionary efforts of Willughby and Ray, Artedi, and Linnaeus as well as the influx of new species of fishes from the tropics. Changing fashions in science also enter the picture, especially the rise of microscopy and microscopic anatomy. Perhaps in terms of the material, questions, and technology, research on chondrichthyans had reached an impasse. There are some exceptions. Bohadsch investigated the structural organization of eggcases. Monro published detailed descriptions accompanied by elegant illustrations of the urogenital anatomy of the oviparous skates.

He also described and illustrated for the first time the spermocysts, spherical aggregations of Sertoli cells and differentiating sperm cells that are a characteristic feature of chondrichthyan spermatogenesis. Descriptions and illustrations of microdissections of the internal anatomy of mid-and late-stage embryos are given. The extensive vitelline circulation and the connection of the yolk sac and the small intestine are documented. External gill filaments, another feature of chondrichthyan development, are described for the first time. The eighteenth century concluded with Bloch's magnum opus (1782-95) which added little to the study of chondrichthyan development other than information on the embryos of the viviparous sawfish Pristis. The study of teleostean development, as previously noted, lagged well behind that of both oviparous and viviparous chondrichthyan fishes, and little was done prior to the mid-late eighteenth century.

To be more accurate, the little that was done was done with viviparous teleosts. Schoenveld first described viviparity in the embryos of the eelpout, Zoarces viviparus, a fish common to the marine waters of northern Europe. Late stage embryos are illustrated. Impressed by the substantial growth of the embryos during gestation, Schoenveld postulated that they obtain nutrients by ingesting ovarian fluid. (In viviparous teleosts, an oviduct or uterus is absent and gestation takes place in the ovary. Cf. Wourms, *et al.*, 1988). Somewhat later, Willughby and Ray apparently observed the development of live Zoarces embryos. Interest in teleost development during the seventeenth and eighteenth centuries seems to have been generated by the viviparity of exotic species discovered during the explorations of the Old and New World Tropics. Anableps, the four-eyed fish, conspicuous because of its size, unusual anatomy, and behaviour in its Central and South American habitats, was described as viviparous as early as 1738. Gronovius was the first to illustrate advanced embryos of Anableps with their characteristic rugose "yolk sac" (actually the abdominal trophoderm). Gronovius may also have been the first to describe viviparity in the clinid, Clinus superciliosus. Bloch described and illustrated the external and internal reproductive anatomy of Anableps, its embryos in situ, and the embryos themselves. He also depicted the adults and embryos of Zoarces and Clinus. One of the first sets of field observations was made by de Alzate — Ramyrez in 1769 who discovered viviparity in a Mexican poeciliid and made microscopic observations of blood circulation in living, near-term embryos.

During this period, the foundations were laid for the study of early development, *i.e.*, fertilization through embryogenesis. In 1763, Jacobi published an extensive letter in which he reported that he had artificially fertilized the eggs of trout and salmon by removing the ripe eggs and adding sperm to them. Although this is the first known written account of artificial fertilization, it is probable that the technique had been in use for some time, for fish culture had been practiced in Europe since the late fifteenth century. Jacobi may have observed the entry of sperm into the egg. He also recorded the formation of double embryos in artificially fertilized eggs, a condition that he attributed to the entry of two spermatozoa. In 1787, Cavolini published a monograph on the "generation" of fishes and crabs. The bulk of this work was devoted to the description of spawning habits, hermaphroditism, fertilization, and contemplations on the nature of development.

However, he did make one observation of considerable significance. In the eggs of the pipefish, he described a structure on the egg which he referred to as the "cicatrix" or in modern terminology, the blastoderm. He identified the blastoderm with the same structure in the chick and assumed that the embryo originated from the blastoderm. The study of teleost development in the eighteenth century was brought to a close with the appearance of Bloch's massive, twelve-volume monograph on the natural history of fishes. This work is one of the linchpins of systematic ichthyology and summarized what was known about fishes up to that time. It contains much information on fish reproduction and development. As previously noted, aspects of the development of viviparous chondrichthyans and teleosts are described and illustrated. Of greater interest are his pioneering studies of the early stages of development of an oviparous teleost. For the first time, he describes and illustrates consecutive stages in development, from post fertilization through hatching to a late larval stage, and provides a time base for the description of developmental stages. He appreciated the eccentric position of the yolk mass relative to the blastoderm. He made observations on the development of the heart, and illustrates the contractile, simple tubular stage of heart development. There were, of course, errors of interpretation, one of which was the identification of the perivitelline fluid of the teleost egg with the albumen layer of the avian egg.

Biotechnology in fish breeding

Gonadotropin releasing hormone (GnRH) is now the best available biotechnological tool for the induced breeding of fish. GnRH is the key regulator and central initiator of reproductive cascade in all vertebrates.It is a decapeptide and was first isolated from pig and ship hypothalami with the ability to induce pituitary release of luteinising hormone (LH) and follicle stimulating hormone (FSH).Since then only one form of GnRH has been identified in most placental mammals including human beings as the sole neuropeptide causing the release of LH and FSH.

However ,in non mammalian species (except guinea pig) twelve GnRH variants have now been structurally elucidated ,among them seven or eight different forms have been isolated from fish species. The most recent GnRH purified and characterized was by Carolsfeld *et al.*(2000) and Robinson et al.(2000).Depending on the structural variant and their biological activities, number of chemical analogues have seen prepared and one of them is salmon GnRH analogue profusely used now in fish breeding and marked commercially under the name of ÒOvaprimÓ throughout the world. The induced breeding of fish is now successfully achieved by development of GnRH technology.

Transgenesis

Transgenesis or transgenics may be defined as the introduction of exogenous gene / DNA into host genome resulting in its stable maintenance, transmission and expression. The technology offers an excellent opportunity for modifying or improving the genetic traits of commercially important fishers, mollusks and crustaceans for aquaculture. The idea of producting transgenic animals became popular when Palmitter et al. (1982) first produced transgenic mouse by introducing metallothionein human growth hormone fusion gene (mT-hGH) into mouse egg, resulting in dramatic increase in growth. This triggered a series of attemptson gene transfer in economically important animals including fish.

The first transgenic fish was produced Zhu et al. (1985) in China, who claimed the transient expression n putative transgenics, although they gave no molecular evidence for the integration of the transgene. The technique has now seen successfully applied to a number of fish species. Dramatic growth enhancement has been shown using this technique especially in salmonids. Some studies have revealed enhancement of growth in adult salmon to an average of 3 Ð 5 times the size of non Ð transgenic controls, with some individuals, especially during the first few months of growth, reaching as much as 10 Ð 30 times the size of the controls.

The introduction of transgenic technique has simultaneously put more emphasis on the need for production of sterile progeny in order to minimize the risk of transgenic stocks mixing in the wild populations. The technical development has expanded the possibilities for producing either sterile fish or those whose reproductive activity can be specifically turned on or off using inducible promoters. This would clearly be of considerable value allowing both optimal growth and controlled reproduction of the transgenic stocks while ensuring that any escaped fish would be unable to breed. An increased resistance of fish to cold temperatures has been another subject of research in fish transgenics for the past several years.

Coldwater temperatures pose a considerable stressor to many fish and few are able to survive water temperatures much below 0-1oC. this is often a major problem in aquaculture in cold climates. Interestingly, some marine teleosts have high levels (10 Ð 25 mg/ml) of serum antifreeze proteins (AFP) or glycoproteins (AFGP) which effectively reduce the freezing temperature by preventing ice-crystal growth. The isolation, characterization and regulation of these antifreeze proteins particularly of the inter flounder Pleuronectas americanus has been the subject of research for a considerable period in Canada.

Consequently, the gene encoding the liver AFP from winter flounder was successfully introduced into the genome of Atlantic salmon where it became integrated into the germ line and then passed onto the off Ð spring F3 where it was expressed specifically in the liver. The introduction of AFPs to gold fish also increased their cold tolerance, to temperatures at which all the control fish died. Similarly, injection or oral administration of AFP to juvenile milkfish or tilapia led to an increase in resistance to a 26 to 13o C. drop in temperature. The development of stocks harbouring this gene would be a major benefit in commercial aquaculture in counties where winter temperatures often border the physiological limits of these species.

The most promising tool for the future of transgenic fish production is undoubtedly in the development of the embryonic stem cell (ESC) technology. There cells are undifferentiated and remain totipotent so they can be manipulated in vitro and subsequently reintroduce into early embryos where they can contribute to the germ line of the host. This would facilitate the genes to be stably introduced or deleted. Although significant progress has been made in several laboratories around the world, there are numerous problems to be resolved before the successful commercialization of the transgenic brood stock for aquaculture.

To realize the full potential of the transgenic fish technology in aquaculture, several important scientific break Ð through are required. There include (i) more efficient technologies for mass gene transfer (ii) targeted gene transfer technologies such as embryonic stem cell gene transfer (iii) suitable promoters to direct the expression of transgenes at optimal levels during the desired developmental stages (iv) identified genes of desireable traits for aquaculture and other applications (v) informations on the physiological, nutritional, immunological and environmental factors that maximize the performance of the transgenics of the transgenics and (vi) safety and environmental impacts of transgenic fish.

Chromosome Engineering

Chromosome sex manipulation techniques to induce polyploidy (triploidy and tetraploidy) and uniparental chromosome inheritance (gynogenesis and androgenesis) have been applied extensively in cultured fish species. These techniques are important in the improvement of fish breeding as they provide a rapid approach for gonadal sterilization, sex control improvement of hybrid viability and clonation. Most vertebrates are diploid meaning that they possess two complete chromosome sets in their somatic cells. Polyploidy individuals possess on or more additional chromosome sets, bringing the total to three in triploids, four in tetraploids and so on. Induced triploidy is widely accepted as the most effective method for producing sterile fish for aquaculture and fisheries management.

The methods used to induce triploids and other types of chromosome set manipulations in fishes and the applications of these biotechnologies to aquaculture and fisheries management are well described. Tetraploid breeding lines are of potential benefit to aquaculture, by providing a convenient way to produce large numbers of sterile triploid fish through simple interploidy crosses between tetraploids and diploids. Although tetraploidy has been induced in many finfish species, the viability of tetraploids was low in most instances.

In teleosts, technique for inducing sterility include exogenous hormone treatment and triploidy induction. The use of hormone treatements, however could be limited by governmental regulation and a lack of consumer acceptance of hormone treated fish products. Triploidy can be induced by exposing eggs to physical or chemical treatment shortly after fertilization to inhibit extrusion of the second polar body triploid fish are expected to be sterile because of the failure of homologous chromosomes to synapse correctly during the first meiotic division.

Methods of triploidy induction in clued exposing fertilized eggs to temperature shock (hot or cold), hydrostatic pressure shock or chemicals such an colchicines, cytochalasin-B or nitrous oxide. Triploid can also be produced by crossing teraploids and diploids. Tetraploid induction involves fertilizing eggs with normal sperm and exposing the diploid sygote for physical or chemical treatment to suppress the first mitotic division. Gynogenesis is the process of animal development with exclusive maternal inheritance.

The production of gynogenetic individuals is of particular interest to fish breeders because a high level of inbreeding can be induced in single generation. Gynogenesis may also be used to produce all Ð female populations in species with female homogamety and to reveal the sex determination mechanisms in fish. It is convenient to use all female gynogenetic progenies (instead of normal bisexual progenies) for sex inversion experiments. Methodologies combining use of induced gynogenesis with hormonal sex inversion have been developed for several aquaculture species. Androgenesis is the process by which would have commercial application in aquaculture. It can also be used in generating homozygous lines of fish and in the recovery of lost genotypes from the crypreserved sperms. Androgenetic individuals have been produced in a few species of cyprinids, cichlids and salmonids.

Biotechnology and fish health management

Disease problem area major constraint for development of aquaculture. biotechnological tools such as molecular diagnostic methods, use of vaccines and immunostimulants are gaining popularity for improving the disease resistance in fish and shellish species world over for viral diseases, avoidance of the pathogen in very important.in this context there is a need to rapid method for detection of the pathogen. Biotechnological tools such as gene probes and polymerase chain reaction (PCR) are showing great potential in this area. Gene probes and PCR based diagnostic methods have developed for a number of pathogens affecting fish and shrimp. In case of finfish aquaculture, number of vaccine against bacteria and viruses have been developed. Some of these have been conventional vaccines consisting of killed microorgansism but new generation of vaccine consisting of protein subunit vaccine genetically engineered organism and DNA vaccine are currently under development.

In the vertebrate system, immunization against disease is a common strategy. However the immune system of shrimp is rather poorly developed, biotechnological tools are helpful for development of molecule, which can stimulate this immune system of shrimp. Recent studies have shown that the non specific defense system can be stimulated using, microbial product such as lipopolysacharides, peptidoglycans or glucans. Among the immunostimulants known to be effective in fish glucan and levamisole enhance phagocytic activities and specific antibody responses.

Cryopreservation of Gametes or Gene Banking

Cryopreservation is a technique, which involve long-term preservation and storage of biological material at a very low temperature usually at -196 C ,the temperature of liquid nitrogen. It is based on the principle that very low temperature tranquilize or immobilize the physiological and biochemical activities of cell, thereby making it possible to keep them viable for very long period. The technology of cryopreservation of fish spermatozoa (milt) has been adopted for animal husbandary.

The first success in preserving fish sperm at low temperature was reported by Blaxter (1953) who fertilizes Herring (Clupea herengus) eggs with frozen thawed semen.The spermatozoa of almost all cultivable fish species has now been cryopreserved. Cryopreservation overcomes problems of male maturing before female, allow selective breeding and stock improvement and enables the conservation One of the emerging requirement for that can be used by breeders for evolving new strains. Most of the plant varieties that has been produced are based on the gene bank collections. Aquatic gene bank however suffers from the fact that at present it is possible to cryopreserve only the male gametes of finfishes and there in no viable technique for finfish eggs and embryos. However , the recent report on the freezing of shrimp embryos. However , the recent report on the freezing of shrimps embryos by subramoniam and newton (1993) and Diwan and kandaswami (1997) look promising. Therefore, it is essential that gene banking of cultivated and cultivable aquatic species be undertaken expeditiously.

Processing and Product Quality and Uses

Biotechnology could have great potential for improving the quality and diversifying the uses of agricultural, forestry and fisheries products. This technology could be used for processing products by both traditional and new methods in order to:

- improve the value, safety and quality of processed and unprocessed foods, such as genetically engineered tomatoes which soften more slowly and which could be processed at lower temperatures, resulting in a higher proportion of solids to liquid in canned tomatoes;
- enhance non-food uses of farm products and thereby increase their market value;
- develop new uses for farm products, especially those often producing surpluses, including energy production from biomass and recycling of wastes and by products. An example of such an application is found in ethanol production from cassava and sugar cane;
- produce algal cells from, for instance, Spirulina or Chlorella in mass culture for proteins;
- produce enzymes on a commercial scale, especially using the microbial system, as substitutes for enzymes isolated from animal or plant sources. For example, chymosin, the enzyme used in the manufacture of cheese, traditionally taken from the cow, can now be made from specially engineered microorganisms. Mass production of amino acids from molasses and other agricultural resources is another example. These amino acids can improve the nutritional value of food and animal feed;
- produce and use new probes for rapid detection and accurate identification of specific microbial pathogens in food, especially diagnostics for staphylococcal enterotoxins, and others; and
- enhance productivity through the application of newly developed processes such as the immobilized enzyme and plant tissue culture systems which are being used to produce useful secondary metabolites. Examples are diosgenin, a female contraceptive commonly used in China and India and shikonin, a natural pigment used for lipstick in Japan.

Biotechnology is being used in a number of countries to characterize the indigenous germplasm variability for therapeutic and biologically important substances.

The use of biotechnology for the removal of toxic and antinutritional factors and for improving the quality of food products is gaining momentum.

These techniques include the elimination or reduction of cyanogenic compounds in cassava and the incorporation of high protein in potato and sweet potato.

Inbreeding

Inbreeding is the mating of relatives. Genetically, all inbreeding does is increase homozygosity in the offspring (this means inbreeding also decreases heterozygosity in the offspring by an equal amount; in this manual, inbreeding will be defined and discussed as it relates to homozygosity).

The increase in homozygosity occurs because related fish share alleles though one or more common ancestors; *i.e.*, the parents may carry a copy of an allele that both inherited from a common ancestor. When relatives mate, the alleles that they share because of their common ancestor(s) can be paired in their offspring. This produces offspring that are more likely to be homozygous at one or more loci.

The mating of unrelated fish also produces offspring that have homozygous genes. Additionally, an inbred fish looks the same as one with no inbreeding; there is no distinguishing mark that separates fish into inbred vs. non-inbred categories. The two types of homozygosity are identical; there is no genetic difference. The only difference between inbred and non-inbred fish is that the homozygosity of each is created in different ways, but this distinction is most important.

Inbred fish are homozygous because they have genes where the two alleles are identical by descent; non-inbred fish are homozygous because they have genes where the two alleles are identical because they are alike in kind. Inbred fish are homozygous at a locus because they inherited identical copies of an allele from both parents that the parents, in turn, inherited from a common ancestor. Non-inbred fish are homozygous at a locus because they just happened to inherit an identical pair of alleles from their parents. This difference in the type of homozygosity is not physical or chemical. The only difference is the paths that the alleles took before they were paired to create the homozygosity; *i.e.*, how the alleles were inherited.

Since there is no difference, why do geneticists make a distinction between the two types of homozygosity? Why is inbreeding of any concern? The reason is that related fish are more alike genetically than unrelated fish. This means that when relatives mate, they produce offspring that have more homozygous loci than is the case when unrelated fish mate.

The mating of relatives produces offspring that tend to be more homozygous than the population average. This increase in homozygosity is called "inbreeding," and the coefficient of inbreeding (F) is a measure of how much more homozygous a fish is than the population average. The coefficient of inbreeding does not measure how many homozygous loci the fish has; it simply quantifies the per cent increase in homozygosity.

On a gene-by-gene basis, F is the probability that the two alleles will be identical by descent.

Since F measures the per cent increase in homozygosity, the same level of inbreeding can produce different amounts of homozygosity in different populations, depending on the level of homozygosity that existed before inbreeding began.

The coefficient of inbreeding merely quantifies the increase in homozygosity that has occurred over a specific time interval (number of generations) due to identity by descent (due to the mating of relatives). Additionally, F is based on probability, so a given value of F is an average value; consequently, a given value of F means more homozygosity will have been produced in some fish, while less will have been produced in others.

The amount of inbreeding that is produced when relatives mate depends on the closeness of the relationship between the parents. Close relatives (*e.g.*, brothers and sisters; parents and offspring) share many alleles inherited from common ancestors, so when they mate, they produce offspring with high levels of inbreeding. Distantly related individuals (*e.g.*, fourth cousins; fifth cousins) share comparatively few alleles from a common ancestor, so when they mate, they produce offspring with comparatively little inbreeding.

Because inbreeding increases homozygosity, it changes genotypic frequencies, increasing the percentage of homozygous genotypes, while shrinking the percentage of heterozygous genotypes. If regular systems of inbreeding are conducted, the breeding programme creates a series of inbred families which do not interbreed. This subdivides a population into numerous lines, which further increases genotypic variance. Figure shows how inbreeding changes genotypic frequency.

While inbreeding changes genotypic frequency, it does not change gene frequency (Figure). Selection, genetic drift, migration, and mutation are the evolutionary and breeding forces that change gene frequency. Inbreeding itself does not alter gene frequency, but by altering genotypic frequency inbreeding can, theoretically, accelerate selection.

During an inbreeding programme, genotypic frequencies can be drastically altered if relatively few inbred families are maintained (a type of selection). When this occurs, gene frequencies will also change.

Uses of Inbreeding

Inbreeding is a breeding programme that can be used to produce superior animal and plant brood stock, and it can also be used to produce genetically improved animals and plants for grow-out. Inbreeding is the breeding programme that is often used to create new breeds or varieties that breed true for "type"; *i.e.,* a particular body conformation or set of qualitative phenotypes. Linebreeding is a form of inbreeding that is used to increase an outstanding animal's contribution to a population. The use of herd bulls is a form of inbreeding that is used to quickly improve a population. Inbreeding can be used as a type of progeny testing to create defect-free animals. Inbreeding can be used to improve response to selection when h^2 is small by combining inbreeding with between-family selection.

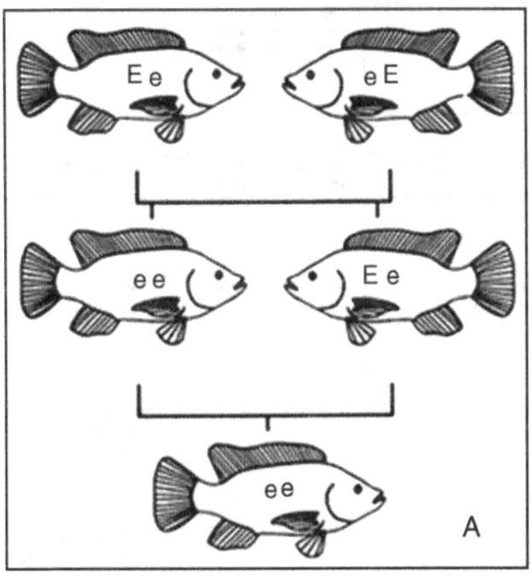

Fig. Fish A is Homozygous By Descent; Fish a is Inbred

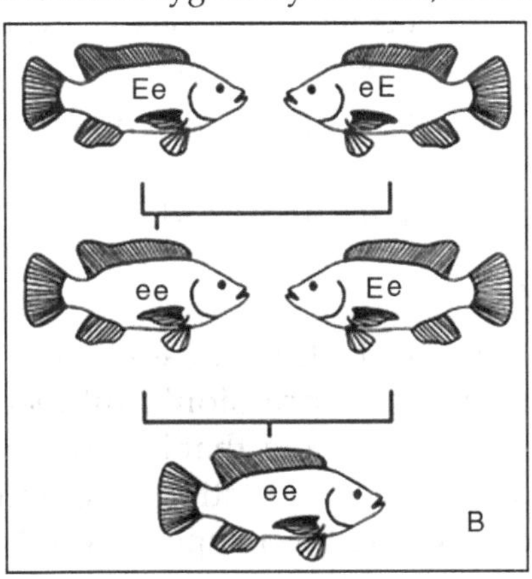

Fig. Fish B is Homozygous in Kind; Fish B is not Inbred

Pedigrees of two fish with identical genotypes (ee), one of which is inbred and one of which is not.

Fish A is inbred because its parents are related (brother and sister); it is homozygous by descent. Fish B is not inbred, because its parents are not related; it is homozygous in kind.

In the P_1 generation, all fish are heterozygotes (Dd). In every generation, the following matings are made: DD × DD; Dd × Dd; dd × dd.

These matings reduce the percentage of heterozygotes and increase the percentage of homozygotes. In the F_3 generation, only 12.5% of the fish are heterozygotes, while 87.5% are homozygotes. Eventually, there will no heterozygotes. Although this mating pattern changes genotypic frequency, the frequencies of the two alleles do not change.

	Genotypic Frequencies			**Gene Frequencies**	
	DD	**Dd**	**dd**	**D**	**d**
P_1					
Frequency	0%	100%	0%	50%	50%
Matings		Dd × Dd			
F_1					
Frequency	25%	50%	25%	50%	50%
Matings	DD×DD	Dd × Dd	dd × dd		
F_2					
Frequency	37.5%	25%	37.5%	50%	50%
Matings	DD × DD	Dd × Dd	dd × dd		
F_3					
Frequency	43.75%	12.5%	43.75%	50%	50%
Matings	DD × DD	Dd × Dd	dd × dd		
$F\infty$					
Frequency	50%	0%	50%	50%	50%

Fig. The Effect of Inbreeding on Genotypic and Gene Frequencies.

Finally, inbreeding is often combined with crossbreeding to increase hybrid vigour. Details about how inbreeding is used and the programmes needed to create inbred lines are described in Chapter.

Inbreeding Depression

As inbreeding increases, it often causes a decrease in productivity which is termed "inbreeding depression". Inbreeding depression is a decrease in growth rate, fecundity, etc. that is observed in the inbred group when it is compared to a control population where there is no inbreeding. The severity of inbreeding depression depends on the level of inbreeding, the phenotype in question, and the population. Inbreeding depression is what gives inbreeding its bad reputation.

There are several explanations for why inbreeding depression occurs, and all contribute to inbreeding depression. The first two are qualitative genetic explanations.

The most common explanation is that inbreeding depression occurs because of the pairing and expression of detrimental recessive alleles. Although the terms "dominant" and "recessive" do not mean that one allele is good and the other undesirable, most of the alleles that produce abnormal phenotypes or that lower viability are recessive. Even though mutation rates are usually only one in 10,000 to one in 100,000 replications per gene, because each fish can produce many gametes (up to 500,000,000 sperm per spawn) and because each fish has tens of thousands of genes in its genome, each fish produces dozens to hundreds of gametes that contain a mutant copy of one or more genes. Most of these mutant alleles are recessive. Since recessive alleles can be expressed only when a fish is homozygous, these mutations tend to accumulate in a population. Consequently, many fish carry "hidden" copies of these mutant alleles in the heterozygous state.

Mutations do not have to be bad; some increase fitness or lead to the evolution of new species by creating new or improved phenotypes. But most mutations lower fitness, because they cause random changes in a phenotype that works and that works well as a result of natural selection. Many mutant alleles produce phenotypes that are so abnormal that they either cause death or reduce viability severely. Others cause only small reductions in viability.

Each fish carries a number of these detrimental recessive alleles, and inbreeding uncovers them. When relatives mate, they produce offspring with an increased level of homozygosity; consequently, some of the detrimental recessive alleles that the parents carry in the unexpressed heterozygous state are paired and expressed in the offspring. Detrimental recessive alleles can also be paired and expressed when unrelated fish mate. The difference is one of magnitude, and that is what causes inbreeding depression.

It has been estimated that each individual carries dozens of mutant alleles that lower viability. Most of these mutant alleles will produce only a small reduction in fitness, but several will produce phenotypes so abnormal that they are lethal or cause premature death. There are many of these unexploded genetic bombs in the gene pool. If two unrelated individuals mate, there is a good chance that they will not carry the same genetic dynamite.

However, when relatives mate, there is a good chance that both parents will carry the same genetic time bombs, because they inherited them from a common ancestor. Even though the probability of producing a defective offspring is the same for each recessive allele (25% if both parents are heterozygotes), if both parents carry a large number of identical defective recessive alleles, the odds of producing defective offspring increase dramatically. The more closely related the parents, the more alike they are genetically, so it is more likely that defective offspring will be created.

Even if two parents do not carry any lethal recessive alleles, they can produce inbred offspring which exhibit inbreeding depression if the parents carry a number of recessive alleles that reduce viability, growth, or other production phenotypes. For example, if two relatives carry 10 recessive alleles, each of which depresses growth by 1%, the pairing and expression of these alleles will depress growth in the inbred offspring. Some offspring produced by these parents will not be homozygous for any of these genes, but most will be homozygous for one or more of these recessive alleles, and growth will be depressed from 1–10% in these offspring.

A second explanation for inbreeding depression is that by decreasing heterozygosity, inbreeding reduces what is called "overdominance". Overdominance occurs when the heterozygous genotype produces a phenotype that is superior to the two homozygous ones. Because inbreeding increases homozygosity, it decreases the number of heterozygous loci, which means that it reduces overdominance. The best known example of overdominance occurs in humans. In many populations, the gene which produces hemoglobin exists in two allelic forms: one produces normal hemoglobin, while the other produces sickle-cell hemoglobin. Individuals who are homozygous normal are susceptible to malaria and get quite ill and can die when infected. Individuals who are homozygous for sickle-cell get sickle-cell anemia and are sickly and often die. The heterozygotes are superior, even though they are slightly anemic, because they are resistant to malaria.

The final explanation for inbreeding depression is a quantitative genetic one. If a quantitative phenotype were controlled solely by V_A, inbreeding would not effect the phenotype, because the types of homozygosity that would be created at each locus would balance over the genome: some homozygous combinations would cause a slight decrease in the phenotypic mean, while other would cause a slight increase.

Since quantitative phenotypes are controlled by hundreds of genes, and since the homozygosity that is created by inbreeding tends to be random, if a quantitative phenotype were solely controlled by V_A, inbreeding-produced homozygosity would neutralize itself, in terms of phenotypic expression. However, few if any quantitative phenotypes are controlled solely by V_A. When V_A controls just 30% of V_P for a particular phenotype, it produces a $h^2 = 0.3$, which is considered to be large.

Since inbreeding increases homozygosity, it also decreases heterozygosity. By doing this, inbreeding affects the interactions between alleles at each locus, which means that it affects V_D, and dominance effects can play a major role in the production of many quantitative phenotypes. When h^2 is small (<0.15), as it often is for viability and fecundity, dominance effects can be quite significant. The creation of excess homozygosity tends to disrupt dominance effects, which tend to be more unidirectional than the additive effects. Inbreeding can be considered to be the opposite of crossbreeding, which is used to exploit V_D. Consequently, if crossbreeding maximizes exploitation of V_D, inbreeding minimizes exploitation, which means that inbreeding makes phenotypes controlled by V_D worse (inbreeding decreases phenotypic mean).

This genetic explanation makes sense, because inbred lines are often used in crossbreeding programmes. The creation of inbred lines is done to produce opposite types of homozygosity that will be maximized when hybrids are created, which is how V_D is exploited. This explanation is further supported by the fact that crossbreeding, the breeding programme that is used to exploit V_D, can also be used to eliminate inbreeding and inbreeding depression.

Inbreeding Studies

Unfortunately, there have been very few inbreeding studies with fish, so there is little specific information about the effects of inbreeding in aquacultured populations. In fact, inbreeding is the least studied aspect of fish genetics. Most of the studies that have been been conducted have been with salmonids, particularly rainbow trout.

Most of the studies are singular experiments that were limited in scope, and many investigated only a single level of inbreeding-usually 25%. Furthermore, most studies looked at the effects of inbreeding in the range of 25–60%. No study has looked at the effects of inbreeding <12.5%. Such studies are needed, because inbreeding experiments with other animals have shown that mild levels of inbreeding can be beneficial.

Because of this, we know relatively little about the effects of inbreeding in fish culture. The studies have shown a few preliminary trends, but we do not know how different levels of inbreeding affect various phenotypes in all important cultured species of fish. We do not know what levels of inbreeding are beneficial and what levels are harmful. Finally, no inbreeding studies have been conducted on some important aquacultured species.

In general, inbreeding studies with fish have shown that inbreeding decreased production phenotypes such as growth rate, fecundity, and survival, while increasing the number of deformed offspring.

The most complete study was a long-term evaluation of the effects of six levels of inbreeding on rainbow trout. Some of the results of this study are summarized in Figures. The results show that the mildest level of inbreeding that was produced (12.5%; this is equal to one generation of half-sib matings) adversely affected per cent hatch and survival, but it increased 77- to 150-day weight. Levels of inbreeding ≥25% severely depressed fecundity, growth, and survival (inbreeding of 25.0%, 37.5%, 50.0%, and 59.4% are produced by one, two, three, and four generations of brother-sister matings).

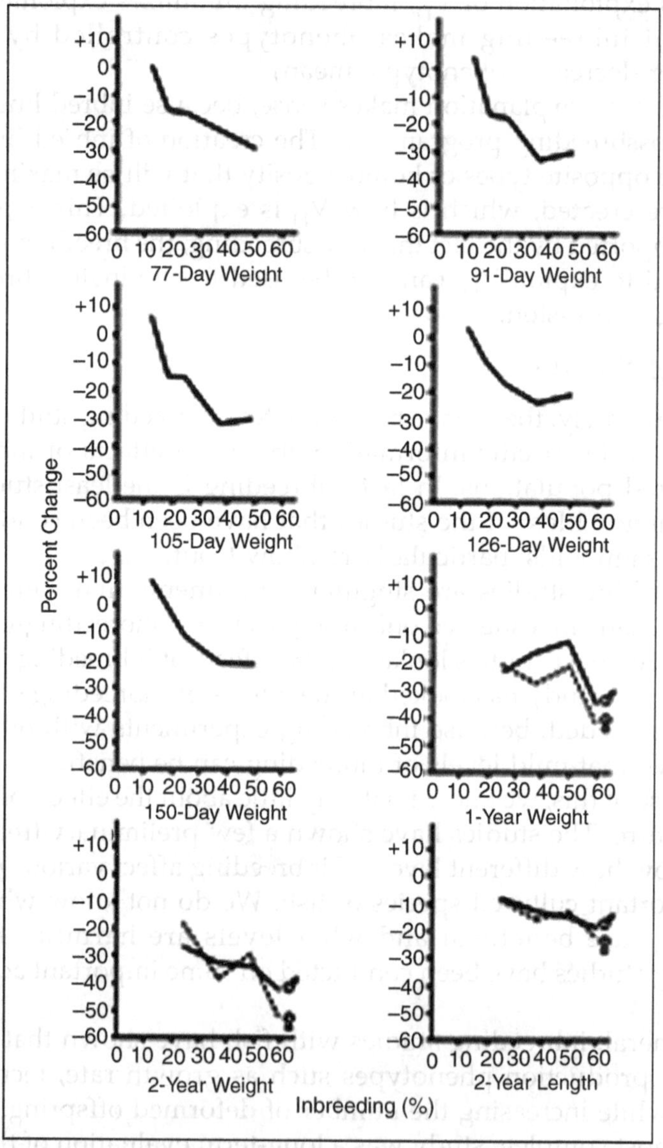

Fig. The Effects of Inbreeding on Length and Weight in Rainbow Trout.

It is not known if these results apply to all species or to all populations of rainbow trout. Studies that have been conducted with other species of fish showed that inbreeding usually decreased means for production phenotypes, although the studies have produced mixed results: for example, inbreeding reduced the return rate of Atlantic salmon, Salmo salar; reduced growth and viability of Mozambique tilapia, Oreochromis mossambicus; increased growth of channel catfish, Ictalurus punctatus, in one study but reduced it in another. The only unambiguous result was that inbreeding produced some abnormal offspring.

It has been suggested that the breeding programmes used by most fish farmers will produce inbreeding of 3–5% per generation. If this occurs, inbreeding depression could begin to affect productivity and profits after only 3 to 5 generations.

The results are presented as per cent change when the inbred group was compared to a control group (inbreeding of 0%); a negative number means the inbred group was worse than the control and exhibited inbreeding depression.

One generation of brother-sister mating produces inbreeding of 25%; two generations produce inbreeding of 37.5%; three generations produce inbreeding of 50%; and four generations produce inbreeding of 59.4%.

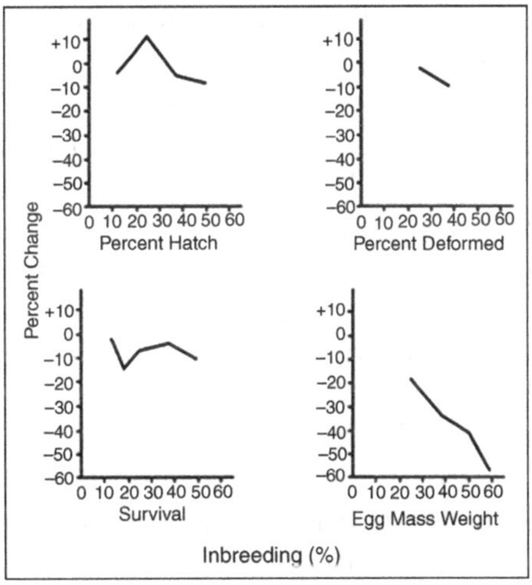

Fig. Effect of Inbreeding on Hatching Rate, Number of Deformed Offspring, Survival, and Fecundity (Egg Mass Weight) in Rainbow Trout.

The results are presented as per cent change when the inbred group was compared to a control group (inbreeding of 0%); a negative number means the inbred group was worse than the control and exhibited inbreeding depression.

Commercialization and Role of the Private Sector

In general, commercialization of agricultural biotechnology is far behind the application of biotechnology in human health and medicine.

In the developed countries, especially in Japan, the private sector plays a major role in the generation, development and commercialization of biotechnology products and techniques. Because of commercial considerations, concentration has been mainly on diagnostics, vaccines, pharmaceuticals and other health-related products. Moreover, because of the eminent involvement of the private sector, marketability of the products and potential return on investment are crucial factors in deciding which products are to be developed and commercialized.

In Japan, the MAFF is subsidizing fundamental technical development projects in the private sector and providing funds for investment in such projects. It is also offering financial assistance to prefectural governments so as to encourage their R&D activities and utilization of the outcome of such activities. The Ministry takes steps to subsidize the research projects of private technical research associations and other similar study programmes. These subsidies are offered mainly through the Ministry's project for the development of biotechnology and other high technology for the food industry. In fiscal 1989, MAFF began such financial assistance programmes as the "Transformation of plant cells through the introduction of small organelles into the cells" and "Genetically manipulated vaccines against animal protozoa diseases" .

In the area of biotechnological R&D activities in the food industry, the Ministry started the project "Development of bioreactor systems for the food industry" in 1984. This project has achieved a number of remarkable results regarding the conversion of saccharides, fat and other substances. Another project, the "Development of technology for improvement of enzyme functions for the food industry" aims to improve the functions of enzymes for food production utilizing recombinant DNA technology. In fiscal 1989, the Ministry began a new programme, "Food production by large-scale, high-density fermentation under ultra-high pressure conditions" with a view to increasing the efficiency of the food production process under ultra-high pressures and to finding food materials with a higher value added.

Two more projects were launched in 1989: the "Utilization of extracts from trees (green sprouts project)" and "Advanced and multipurpose utilization of unused marine resources (marine-frontier project)". The objective of these projects is to develop the technique for extracting useful substances from trees and marine organisms and utilizing them for foods, perfumes and medicines, thereby helping further growth of forestry and fishery industries.

The Bio-Oriented Technology Research Advancement Institution (BRAIN) was founded in October 1986 by the joint investment of the government and the private sector. In an effort to encourage the R&D

activities of the industries related to agriculture, forestry, fishery and food production, this institution is investing in joint technical development corporations and is providing no interest-bearing loans to the technical research projects of businesses on certain conditions.

During three years from fiscal 1986 to 1988, BRAIN invested in 14 study projects and offered loans to 70 projects. In fiscal 1989, the Institution invested in, among others, the Institute of Aquacultural Technology of Cold Water Fishes, Wakayama Agro-biological Research Centre and the Japan Turf Grass Co. The first is engaged in the development of aquaculture systems of cold water high-class fishes, the second in R&D activities for rationalized process of tangerine juice production and the third in the breeding of better turfs.

In the area of biotechnology-related R&D in the private sector, financing systems through the Japan Development Bank, the Agriculture, Forestry and Fishery Finance Corporation and other similar government organizations were reinforced to turn the results of these R&D activities to practical use. It is hoped that as biotechnology is increasingly introduced to industries, these financial measures will be utilized more in the coming years.

Prefectures are becoming more and more interested in R&D activities on biotechnology and the practical use of the outcome of such activities. This is evident from the fact that almost all prefectures throughout Japan have a council on this technology composed of scholars and other people of expertise.

MAFF is offering to prefectures a variety of subsidies in order to diffuse the mass production technology of virus-free seedlings of vegetables, ornamental plants and fruit trees and the transplanting technology of the embryos of beef cattle. In fiscal 1989, the Ministry started the "Project for establishing embryo supply centres" to ensure a stable supply of embryos of beef cattle. This project is improving the raising and controlling facilities of cattle from which such embryos are taken.

Since 1986, MAFF has also been implementing the "Project for assisting local R&D activities in biotechnology". This project offers financial assistance to the R&D activities of prefectural research institutes concerning the breeding and utilization technology of biological resources. Other biotechnology-related activities of the Ministry include regional biotechnology meetings held in seven locations across the country with the object of exchanging information on R&D. The Ministry's new initiative in fiscal 1989 included surveys and studies for examining the direction of the practical use of biotechnology in each region. The Tsukuba Bioscience Hall created in 1989 provides a forum for collaboration among trainees from industries, universities and the government and international exchange programmes. Furthermore, to foster links among genetic resources and biotechnology, the Ministry has a strong national genebank and germplasm conservation programme .

Of the developed countries in the region, the involvement of the private sector in biotechnology, especially plant biotechnology, is rather limited in New Zealand. Very little research on plant biotechnology is being carried out by private companies. A shortage of venture capital, low taxation incentives and tough competition from large overseas firms make such ventures unattractive.

Several small companies are engaged in either product formulation, or manufacture of biological control agents. Larger international companies (*e.g.* Monsanto, MPI Koln, PGS Ghent, ICI) and overseas research institutions (UCLA, California; John Innes Institute, United Kingdom; Christian Albrecht University, Germany) have provided personnel, facilities and some supportive funding on collaborative projects. Producer Boards (*e.g.* New Zealand Kiwifruit Marketing Board) also support research through joint funding of government programmes .

Many New Zealand research programmes on plant biotechnology are not mature enough to have reached the stage of successful commercialization. In addition, some programmes are commercially sensitive, particularly where research is collaborative with private companies. Examples of plant biotechnology research that have been, or are in the process of being commercialized in New Zealand include :

- Establishment of a commercial tissue culture laboratory for clonal Forestry with Radiata Pine (FRI).
- Use of immuno-molecular techniques for developing test assay kits and for analysis of gene expression (DSIR, Fruit and Trees).
- Plant variety rights have been granted for four new plant varieties.

Future commercialization is likely to be through the release of improved crop varieties, development of new cultivars and the licensing of technologies that have been developed.

In the developing countries, in vitro culture for micropropagation of elite and disease-free materials, production and distribution of efficient nitrogen-fixing microbes, diagnostic kits and monoclonal antibody-based vaccines are the main areas that have been commercialized to varying degrees in different countries.

In developing countries, the role of the private sector in modern biotechnology has been rather limited. This is attributed partly to: (i) vague government policies regarding the private sector; (ii) unclear policies or no policies/views on patents and intellectual property rights; (iii) poor links between public and private sectors; (iv) low purchasing power for new products that are usually highly priced and are out of reach of the majority of resource-poor farmers and low-income consumers; and (v) inadequate local expertise and infrastructure and R&D support to new biotechnology.

In India, biotechnology industries have recently been growing in the private sector. Hindustan Lever Ltd, an Indian subsidiary of the multinational Unilever, the Tata Energy Research Institute (TERI) and Southern Petrochemical Industries are some of the leading biotechnology companies in the country. These have considerable in-house R&D bases active in the application of new technology to agribusiness, such as enzyme technology for high-value chemicals and quality edible oil by using genetically modified bacteria; protoplast fusion in yeast for efficient fermentation and production of edible oil from biomass; tissue culture of plantation and ornamental crops; biological nitrogen fixation; organic components for higher rate of photosynthesis; big-insecticides; plant growth promoters; biomass processing for animal feeds; and hybrid seeds. Some small and medium-size companies are also initiating biotechnology activities but, with less R&D capacity, are more wary of what they consider to be a higher-risk technology.

The private sector in India is particularly active in tissue culture products. There are about 50 private sector companies such as Indo-American Hybrid Seeds, A.V. Thomas and Co., Unicorn Biotek, the three companies and others that are comprehensively involved in commercial in vitro production of planting materials of several horticultural and plantation crops such as cardamom, coffee, tea, oilpalm, orchids, strawberry, roses, Spatiphyllum, banana, lily, Gerbera, etc. both for domestic and export markets. These companies have established effective links with the public sector .

The recent liberalization of trade, including that of certified seed, coupled with the greater priority given to science and technology and the promotion of private sector and public-private sector links, should encourage increased participation of the industry in agricultural biotechnology R&D. Financial incentives are available for supporting the growth of indigenous biotech industries. However, venture capital is still lacking. Several multinationals, *e.g.* Seedtec International, Cargil Inc., Dehlgien Inc., Northrup King and Sandoz (through their local subsidiaries) have been shortlisted by the Indian Government for production of hybrid seeds which may involve biotechnologically produced seeds in collaboration with Indian companies. To take advantage of the availability of expert personnel in biotechnology at relatively low costs, a few multinational companies like Astra AB (Sweden) are setting up a biotechnology R&D centre in India. Recently, Pro-Agro, an Indian seed company, established links with Plant Genetic Systems, a leading Belgium-based plant biotechnology company, for the production and distribution of hybrid seeds.

In Indonesia, there is relatively little direct involvement or sponsorship of research in government institutions at present. Some companies are investigating improved planting materials. A commercial production system has been started for improved oil palm plantlets, with a target of 1 million plants per year in 1993, at the Research Institute for Oil Palm at Marihat. Animal vaccines are produced by Pus Vetima and human vaccines by Perum BioFarma, in government-owned companies, which also undertake some biotechnology research.

In the Republic of Korea, the entrepreneurship drive led the biotechnology sector to be organized in 1982 under the Korean Genetic Engineering Research Association (KOGERA). At present more than 20 private companies are participating in this Association. The government provides the members of KOGERA with subsidy as the seed money. In 1980, the public sector provided 70 percent of the R&D support to the biotech industry. Six years later, the situation was reversed; in 1986 the public sector provided only 28 percent of the R&D cost whereas the industry provided 72 percent of the cost. This shows the increasing success and confidence of the private sector. The private companies are mostly involved in fermentation and pharmaceutical matters and have very little interest in agricultural biotechnology. Most of the companies are using biotechnology for the production of antibiotics, insulin, new vaccines and interferons.

The Association performs an important role in the exchange of biotechnology information and also promotes relevant research. MOST provided a special fund for 122 projects in the field of gene manipulation and recombinant DNA technology in 1990. Of these national projects, 32 projects were directly connected with crop improvement . The Korean Science Foundation (KSF) has contracted research projects to universities relating to biotechnology in the fields of medical, pharmaceutical, and agricultural sciences.

The entire biotechnology work in agriculture in the Republic of Korea is coordinated by the Agricultural Biotechnology Research Council comprising researchers from national institutes and universities and representatives of industries and the private sector. A few private companies involved in agricultural biotechnology are using the new technology for the production of virus-free potato planting material, and through the anther culture technique are producing haploids for heterosis breeding in Chinese cabbage and rice. The private sector, besides having strengthened its own R&D system, also provides assistance to universities for undertaking specific biotechnological research. The increase in personnel and funds for biotechnological research and development in the private sector in the Republic of Korea.

Intellectual Property Rights

Two Intellectual Property Rights (IPR) mechanisms namely, patents and plant breeders' rights are most important for agricultural biotechnology. A patent is a right granted by the government to inventors to exclude others from imitating, manufacturing, using or selling a specific invention for commercial use during a certain period, usually 17-20 years.

The patent holder, in turn, is obliged to disclose the invention to the public. Plant Breeders' Rights (PBR) are rights granted by the government to plant breeders to exclude others from producing or commercializing material of a specific plant variety for a period of about 15 to 20 years.

Both under IPR and PBR, use of the protected matter is restrictive, being highly restrictive under IPR. There is, however, provision for research exemption under IPR which allows others to study the protected subject-matter without reproducing or multiplying it for commercial purposes.

Under PBR, there are provisions for breeders' exemption and farmers' privilege. But, in 1991, UPOV tightened PBR by eliminating the breeders' exemption for an essentially derived variety-a variety predominantly derived from another (initial) variety which retains the expression of the essential characteristics from the genotype of the initial variety.

Now, as per the 1991 legislation, a breeder who inserts a single new disease-resistance gene into a PBR-protected variety will have to obtain permission from the holder of the original rights before marketing the new variety. In 1991 UPOV also attempted to tighten up farmers' privilege, *i.e.* freedom to farmers' freedom to use their own harvested material of protected varieties for the next crop on their farm, but because of a lack of consensus among UPOV members it was left to the national governments whether to permit farmers to reuse the seed of a BPR-protected variety for further crop cycles on their own holdings.

The status of development and adoption of IPR and PBR varies widely from country to country in the region. The three developed countries (Australia, Japan and New Zealand) are all members of UPOV and have explicit IPR provisions for biotechnological products and procedures. None of the developing countries are members of UPOV and have mostly been opposed to enforcement of any form of IPR, particularly patents.

However, lately, under various kinds of bilateral pressures, such as the special 301 Provisions of the US Omnibus Trade Act of 1988, as well as multilateral pressures, and after having signed the successfully negotiated GATT agreements in December 1993, whether to introduce legislation covering proprietary protection for biotechnology products and procedures, and/or how rigid the protection system should be, are under active debate in several developing countries of the region.

Those in favour of having the legislation argue that an effective IPR system would: (i) improve the access of the country to new technology from abroad; (ii) stimulate private sectors; and (iii) encourage foreign investment.

Those against it argue that in the context of the developing countries it will: (i) suppress in-country innovation and local industries and production systems; (ii) increase the dependence on foreign countries even for materials and procedures that are vital for the people of the importing country; (iii) increase prices of even commonly used products and services; and (iv) restrict information flow and knowledge growth.

The effects of IPR on the development of biotechnology in developing countries are not clearly understood and should be studied systematically, covering countries at various levels of development and of varying sizes and in the context of different industries, such as pharmaceuticals, agriculture, food industries, environmental protection, etc.

Thailand adopted a new patent act (1992) that protects biotechnology inventions after the United States Government had withdrawn Thailand's trade benefits under the General System of Preferences. Under a similar pressure, when the renewal of the Sino-American Science and Technology Agreement was postponed by the United States Government, China revised its patent law effective as of January 1993, which protects biotechnology inventions, but excludes plant and animal varieties.

In India, the matter is hotly debated. Despite the pressure from the United States Government by postponing the Indo-US Vaccine Action Programme and the Indo-US Science and Technology Initiative, the Indian Government has not yet strengthened its IPR legislation. However, given these pressures, the changed and liberated seed policy of the government and increasing pressure from the private seed companies within the country, there is a change in the mood of the government, although strong opposition persists. India signed the GATT agreement in December 1993. Its 1994 Spring Session of the Parliament held a special debate on the issue, and the opposition parties continue to agitate. It appears that a consensus may emerge to evolve an IPR system that will not promote inequity, jobless growth, and ecological degradation.

The government is currently preparing an effective sui generis UPOV PBR, explicitly emphasizing breeders' exemption and a strong farmers' privilege.

In addition, the system would provide for farmers' rights and include mechanisms for the operation of these rights.

As regards other developing Asian countries with considerable programmes in agricultural biotechnology, no patent protection systems exist either in Malaysia or the Philippines. However, Indonesia is preparing a new IPR legislation.

It can be seen from the above that, under various kinds of pressures, from outside or within, and as biotechnological R&D efforts are being intensified, several developing countries are moving towards restructuring or strengthening their IPR systems.

What is important at this stage is that the countries strike a rational balance between the positive and negative aspects of the IPR system and tailor their legislations to their needs, opportunities and aspirations. In countries where the local technology base is weak, an IPR system may contribute little to the indigenous technology capacity.

However, if the country is now able to pursue cutting-edge biotechnology and produce innovative and useful products and processes at competitive prices, the relevance of an appropriate IPR would increases. The countries must therefore be able to analyse the various pros and cons and short-and long-term socio-economic and ecoenvironmental advantages and disadvantages of instituting a particular IPR system. International organizations, such as FAO, should assist the countries in this task.

Regulatory Aspects

Biosafety

R&D biotechnology is heavily influenced by two regulatory issues, namely biosafety and intellectual property rights. Biosafety aspects include the policies and procedures needed to ensure the environmentally safe development and application of biotechnologies. An effective biosafety system to regulate release and use of Genetically Modified Organisms (GMOs) and to address concerns about potential risks to public health and environment must be in place in all countries involved in biotechnology. Biosafety regulations are necessary also to facilitate technology transfer. Comprehensive biosafety guidelines and measures exist in the developed countries, but the situation is far from satisfactory in most developing countries, although they are aware of the need and are in the process of formulating national frameworks, procedures and capabilities for addressing biosafety issues.

In Australia, biosafety guidelines/regulations have been evolving since 1975, when the Academy of Science Committee on Recombinant DNA (ASCORD) issued its first guidelines. In 1981, as per the Commonwealth resolution, the Recombinant DNA Monitoring Committee (RDMC) was established. In 1987, the RDMC published "Procedures for the Assessment of the Planned Release of Recombinant DNA Organisms". The RDMC was further transformed into the Genetic Manipulation Advisory Committee (GMAC) to include the surveillance of any genetic manipulation procedure resulting in an organism containing foreign DNA that was unlikely to be formed by conventional breeding practices. The GMAC emphasizes that the focus of the assessment should be directed towards ensuring that the release does not cause harm to people, livestock, plants or the natural environment, and not primarily be directed towards considering how the organism was constructed.

GMAC adopted the following procedure :

1. Approval of the concerned Institutional Biosafety Committee (IBC) is obtained.
2. The IBC sends the proposal to GMAC.
3. GMAC, according to its Procedures for the Assignment of the Planned Release of Recombinant DNA Organisms, in consultation with its Scientific and Planned Release Committees, analyses and evaluates the proposal, especially the genetic aspects and any possible environmental impact associated with the construct.
4. The above assessment report, especially pointing out hazards, if any, is sent to the proposer and to the agency that has legal jurisdiction over the release of that particular class of organism.
5. The responsible agency then makes the decision whether to release or not to release the organism based on the report from GMAC and other related considerations."

In Japan, five guidelines have been established for the regulation of research of DNA organisms and the safe promotion of experiments since 1979. Appropriate regulations are considered effective in minimizing the potential risks of a new technology to the environment and human health, and in ensuring positive public perception. The purpose of MAFF's "Guidelines for the Application of Recombinant DNA Organisms in Agriculture, Forestry,

Fisheries and the Food Industry and Other Related Industries", is to promote the safe progress of agro-industries by: (i) defining general principles for the appropriate application of rDNA organisms; and (ii) ensuring safety in the use of rDNA organisms . The guidelines are based on OECD guidelines.

Applications for field trials of rDNA organisms are to be approved by MAFF through biosafety assessments conducted by the authority committee under MAFF. The first application for field tests using transgenic tomato plants harbouring TMV coat protein genes was filed in late 1990. The application was based on three years of greenhouse experimentation. As the biosafety review of this application was completed, the first field trials were approved in 1991 and since then many field tests have been approved.

In New Zealand, an Advisory Committee on Novel Genetic Techniques (ACNGT) was set up in 1978 under the terms of a Cabinet resolution. All persons working in the public sector (universities, government departments, etc.) are required to inform ACNGT of their intention to conduct experiments that are defined as "any experiment involving novel genetic techniques that is properly contained and that can be terminated at any point without loss of containment" .

In 1988 a Genetically Modified Organism Interim Assessment Group (JAG) was established under the Environment Act. The principal role of IAG is to assess all proposals to field test or release GMOs and in so doing provide effective opportunity for public involvement in the assessment process.

Private sector researchers are not bound by IAG but are encouraged to use it. In 1989 the New Zealand Government announced that a new agency, the Hazard Control Commission, would be established. This Commission, in addition to responsibilities with respect to hazardous substances, is responsible for assessing and licensing all genetic manipulation work in New Zealand and approving applications to import or release new organisms.

Among the developing countries, comprehensive national biosafety systems/guidelines have been set up in China, India, the Philippines and the Republic of Korea. Most of these guidelines are based on multitier approach, modelled on the lines of OECD and NIH and NAS (United States) biosafety procedures and guidelines for handling and release of GMOs.

In India, the Recombinant DNA Advisory Committee of the Department of Biotechnology (DBT, 1990) prepared a set of Recombinant DNA Safety Guidelines that cover all areas of research and large-scale operations involving genetically engineered organisms, excluding humans. The institutional mechanism for implementing the guidelines consists of:

- the Recombinant DNA Advisory Committee, under DBT, to formulate and update biosafety guidelines;
- institutional biosafety committees located at all centres engaged in genetic engineering research and production activities;
- the Review Committee on Genetic Manipulation, under DBT, to guide the institutional biosafety committees;
- the Genetic Engineering Approval Committee, functioning under the Department of Environment, to review and approve activities

involving the large-scale use of genetically engineered organisms and their products in research and development, industrial production, environmental release, and field applications.

In the Philippines, the National Committee on Biosafety and Biosafety Guidelines is a working group composed of representatives of the Department of Agriculture, the Department of Science and Technology, the University of the Philippines, and the International Rice Research Institute .

It has prepared biosafety guidelines that include a provision for case-by-case review of proposed releases of genetically modified organisms. A national Biosafety Committee was established in 1990 to oversee the compliance of policies and guidelines in public and private institutions. A network of 39 Institutional Biosafety Committees was created to ensure the implementation of the guidelines at the institutional level. The National Biosafety Committee coordinates with other national agencies involved in regulations, such as the quarantine services and the environmental management bureau. To date, the National Biosafety Committee has approved 15 applications for the importation of, among other things, transgenic cotton tissue and transgenic strains of rice.

Biology of Artificial Propagation

The artificial propagation means a series of processes from sexual maturity, ovulation, spawning upto incubation of fry under control by man. Gonads are also termed genital glands. The gonad of female fish is ovary, that of male fish testis. The eggs and sperms develop into maturity respectively in ovaries and testes when fish get to the age of sexual maturity. The gonads have cyclic variance in development, and in the process of it, they are controlled by internal endocrine and nervous systems and external ecological conditions. The two factors are associated and restricted each other.

Structure of ovary and ovum, stage division of ovary

The ovaries are in pairs, symmetric and sac-like. Their walls are formed by connective tissues and smooth muscles. The inner wall of ovary protrudes above septum, named as ovum-producing plate, where ova are turned out. Following maturity, the ova break follicular membranes and drop into the ovary cavity, the end of which adjoins a short oviduct with an opening to the exterior of body. There are blood vessels and nerve branches on the ovary tissue.

Just like other animals' eggs, fish's ova possess the common features of body cells, consisting of cytoplasm (ovum cytoplasm), cell nucleus (ovum nucleus) and cell membrane (egg membrane).

At the initial stage of development, the sphere-shaped egg cells only have nucleus and cytoplasm. In further development, the egg will accumulate egg yolk materials by and by. The amount of egg yolk is much larger than that of egg cytoplasm. Polarization appears in their distribution, thus the egg is called polarized yolk egg thereby.

The egg is topped by the so-called animal pole. Except for the yolk, most of the egg constituents such as cytoplasm and nucleus are concentrated near this pole. The yolk concentration increases towards the opposite, vegetal pole because of its larger granule and specific gravity.

After fertilization, the first mitotic division of the cell cytoplasm and nucleus will take place and embryonic development will begin. Egg yolk neither takes part in nor hinders the process of mitotic division. Egg yolk contains protein, fat, glycogen and vitamin etc., which are the nutritional materials for embryonic development. The nucleus consists of nuclear membrane, nucleole, nuclear fluid and chromosome. The nucleus is sphere-shaped but often leaf-shaped. Its function is to maintain genetic materials (DNA) and cell metabolism, passing the genetic material on to one generation after another.

The cell membrane is a membrane structure covering the exterior of egg cell, that can be divided into 3 kinds in accordance with the materials. Primary cell membrane or yolk membrane is made of cell's own plasm, with lots of radiant duct-like pores, so-called radiant belt, which are helpful for the absorption of nutrients and the discharge of metabolic wastes. Secondary cell membrane or chorion is secreted by follicle cells in ovary. The secondary cell membrane of fish is normally adhesive.

It is very common in adhesive eggs of Common carp, Crucian carp and Wuchang fish etc.. Tertiary cell membrane is composed of the secretion from glands in oviduct, such as gummy cell membranes of frog and cuttlefish. Fish do not have all the three above-mentioned membranes. Silver carp, Bighead, Grass carp and Black carp have just the primary cell membrane and Common carp, Crucian carp and Wuchang fish primary and secondary membranes.

The development stages of the ovary

The development stages of ovary maturity can be judged by visual observation or histologic survey. The development of fish gonad may be generally divided into six stages according to its appearance, colour, size, weight, and the distribution of blood vessels and the maturity of ova. Nowadays, the classification of the development stages of ovary adopted by each country is not identical.

Five stages are admitted in India, Japan and the United Stages of America. Several countries prefer seven stages, whereas China does six. Set Silver carp as an example:

Stage I Ovary

By visual observation, the gonads are located at the lower part of air bladder, closely attached to the coelomic membrane, lineal in shape, transparent, flesh-white in colour and it's impossible to distinguish them into different sexes by the naked eye.

Tissue slice: cells tiny, diameter 12—22 um; nucleus rather big, occupying more than half of oocyte's diameter; very few nucleoli in the centre of the nucleus.

Stage II Ovary

By visual observation, ribbon-shaped, flesh-white in colour and semitransparent. It is impossible to distinguish ova one after another by the naked eye, and yet small eggs are visible when examined under a magnifyingglass; petaline-shaped when in fixation. It is possible to distinguish sexes. The maturity coefficient is about 1—2%.

Tissue slice: cell in multiangular shape or sphere-shape with a diameter of 90—300 um. A thin layer of ribbon-shaped follicles surrounds cell membrane surface. The nucleoli closely attach to the nuclear membrane

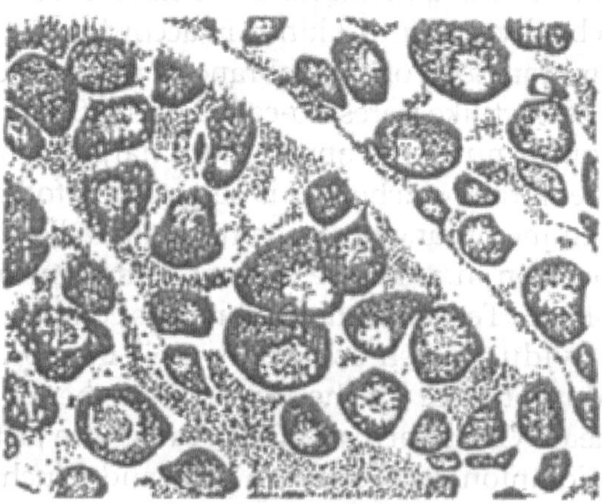

Fig. : Tissue Slice of Stage II Ovary of Silver Carp

Stage III Ovary

By visual observation, the capacity of the ovary has become conspicuously enlarged. It is due to the appearance of melanotic pigment so that it turns greenish grey in colour. Eggs are visible with the naked eye but not easily separable.

The distribution of blood vessels is clear. The maturity coefficient of ovary is 3—6%. Tissue slice: The follicular membrane in oocyte is bilayer. The egg yolk begins to form. One or two layers of vacuoles appear on the edge of the cell. The cell is 250—500um in diameter. The nucleus in the centre is irregular or oval-shaped. Greater part of nucleoli are distributed on the edge of nuclear membrane and small part scattered in the centre.

Stage IV Ovary

By visual observation, it is long and sac-lake, occupying 1/3—1/2 of coelomic cavity. Eggs are plump and greenish grey or light yellow in colour and easy to separate. Ovary is fully distributed with blood vessels. The maturity coefficient is 12—22%.

Tissue slice: egg yolk granules fill almost all the space outside nucleus with only a little cytoplasm spreading around nucleus and near egg membrane. The egg is enlarged to 800—1580 um in diameter. Nucleus edge looks like waves in shape with a few nucleoli inserted in the troughs and most of nucleoli moving to the centre.

This stage can be further separated into 3 stages in the light of occyte's diameter and nucleus location. Early stage IV: egg diameter about 800 um, nucleus in the centre; middle stageIV: egg diameter about 1000 um nucleus in the centre or inclined to move to animal pole; late stage IV: egg diameter about 1580 um, nucleus having moved to animal pole, which is so-called polarized phenomenon.

Lots of experiments and practices have proved that no matured eggs can be obtained by inducement as the oocytes of silver carp, Bighead, Grass carp and Black carp are at early stage IV. Only in case of middle and late stage IV when the nucleus is eccentric or polarized, matured eggs can be acquired, and then the artificial estrualization will succeed. These stages could last as long as 1—2 months, even 3 months for some cases, providing that proper ecological conditions for spawning are not available during that period of time or no artificial propagation is performed.

Stage V Ovary

By visual ovservation quantities of oocytes come out from follicular membranes into ovary cavity, and eggs are in a state of flowing. The ovary and the belly are very soft. The eggs would flow out from cloacal opening by way of a slight pressure upon belly.

Tissue slice: the yolk granules start to fuse into a lump. The cytoplasm and the nucleus have already moved to animal pole. The nucleoli concentre in the centre of the nucleus, whereas the nuclear membrane has dissolved through perforation. The nucleus looks transparent.

When oocytes proceed to their maturity, the epithelial cells of follicles secrete a sort of substance which can dissolve and absorb tissues between the follicular and the egg membranes; thus the eggs can easily be released out from follicles thereby. The eggs are in a state of flowing in the saclike.

ovary. This process is called ovulation. The eggs, which are in a state of flowing, are released from the fish body, this process is termed spawning. The oocytes proceed very fast from stage IV to maturity (stage V).

In nature, the process may be fulfilled in 20—40 hours after the rising of water in the rivers. Under artificial estrualization, it can generally be fulfilled within 10 to 20 hours or even in several hours. If the follicles discharged inmatured eggs ahead of time, the rate of fertilization would be affected. In another case, if the follicles could not release matured eggs in time, the eggs might be suffocated or degenerate by virtue of overripe and the rate of fertilization would be certainly affected too. Even if part of these eggs might become fertilized, the embryos could not develop normally. In other words, the success of either natural spawning or artificial insemination depends upon knowing exactly the maturity stage and the spawning time of a fish. If the matured eggs missed the chance of being released, the oocytes would degenerate and be absorbed.

Stage VI Ovary

By visual observation, the major portion of eggs has already been laid. There are still some phase-4 oocytes which have remained in the ovary. The ovary is slack and its size has been reduced conspicuously. Blood vessels have become enlarged with lump-shaped extravasated blood. Tissue slice: After ovulation, there are lots of follicular membranes and undischarged mature eggs in ovary. The undischarged eggs will soon degenerate and be absorbed, forming a semi-transparent and irregular structure, orangeyellow in colour. Besides there are still many interim oocytes in it.

Future Prospects and Strategies

Alleviating the widespread food gap and attaining food security or food reliance and nutritional adequacy should be the highest national priority in almost all African countries. In order to achieve this foremost goal, production, productivity and sustainability of food and agriculture must be greatly enhanced through the generation, development and transfer of appropriate technologies.

While the potential of currently available technologies should be rationally harnessed, new and emerging technologies such as biotechnologies, should be used in conjunction with, not as substitute for, existing conventional technologies.

Biotechnology is offering innovative possibilities for increasing crops, livestock, forestry and fisheries production and for protecting the environment by the reduced use of agrochemicals, conservation of genetic resources, bioremediation and recycling of wastes. In vitro culture techniques for micropropagation, freeing planting materials of viruses and other pathogens, the production of secondary metabolites and germplasm conservation are being widely used.

Genetic mapping of major tropical crops as an aid to conventional plant breeding programmes; plant virus resistance by genetic engineering of the host plant; novel biocontrol agents for pest control to reduce pesticide use; and the use of efficient and new biofertilizing agents are now being increasingly used mostly in developed and in a few developing countries. Embryo transfer techniques and the use of diagnostic kits for detection of diseases both in plants and animals, and the use of highly specific and effective vaccines are proving of great value to animal production and health.

Biotechnologically designed new feeds, hormones and disease control agents are already favourably impacting fish production. But, in Africa, these technologies and products are being used only rarely, in a limited way and in an ad hoc manner. The slow and poor development of modern biotechnology in Africa may be ascribed to: (i) lack of appropriate policies, strategies, plans and programmes and the absence of political will in general; (ii) inadequacy of human resources, paucity of financial investments, poor research and laboratory infrastructure, absence of priority-setting mechanisms and insufficiency of mechanisms to analyse global market and technology trends; and (iii) negligible involvement of the private sector and absence of effective regulatory measures.

In view of the above, the African strategy for biotechnological development should include both the provision of direct support for biotechnological research, product development, technology assessment and transfer and the creation and fostering of a climate where biotechnology can develop in a balanced manner. This will largely depend on governmental and public awareness, political will, policies and programmes for building national capabilities and strengthening/ developing regional and international cooperation.

Creative partnerships between public and private sector interests are critical in establishing a competitive strategy in biotechnology. While developing institutional mechanisms best suited to a given country, it is essential to establish effective interdisciplinary collaboration among molecular biologists, breeders, agronomists and plant protection specialists and to forge and menage linkages among researchers, extensionists, farmers and other users of bioproducts and techniques.

It may be generally preferable to build the new initiatives on traditional strengths in agricultural, forestry and fishery research, rather than creating totally new institutes. It should be kept in mind that biotechnology in crop and livestock production will be applied through the creation of new genotypes possessing novel characteristics, and this is possible only when conventional breeding and production systems are already in place.

Development of trained personnel should be a high priority in Africa. The important areas of training that are required for a successful biotechnology programme are microbiology, plant biology, zoology, cell biology, virology, genetics, biochemistry and cell and molecular biology.

In most of sub-Saharan Africa, the people trained in these areas are in very short supply. Even biology and tissue culture training is not in the currriculum of most tertiary education institutions. Very few universities offer training with adequate laboratory facilities in molecular biology and recombinant DNA technology.

Those that are trying to offer such training are constrained by the lack of equipment and staff. In recent years, however, some donor-supported formal training programmes in biotechnology have been developed. For instance, the M.Sc. in biotechnology at the University of Zimbabwe was assisted by SAREC (Sweden) and DGIS/Ministry of Foreign Affairs (the Netherlands).

The donors provided the funding for both the teaching/research equipment as well as funds to buy supplies and to cover the cost of bringing Dutch and Swedish biotechnology specialists to Zimbabwe to give special lectures in areas in which there was no local competence. Students from Zimbabwe have also been admitted to universities in the donor countries for Ph.D. degrees. Such arrangements exist and/or are being developed in several other African countries.

The actual and potential effects of biotechnology development are not always positive. In the African context, the negative effect is the displacement of crops at present grown for export in Africa by biotechnologically designed "industrially" produced analogous in industrialized countries.

African countries should be aware of such developments and diversify their products or increase quality and cost-effectiveness of their products not only to maintain the international market share but also to enhance or open new opportunities.

FAO and other concerned international systems should assist African countries in developing early warning systems whereby the potential relative substitution effects could be monitored and strategic adjustments brought about in time to avert damage.

In industrialized countries, where biotechnology is essentially in the hands of the private sector, the work is focused on areas and commodities that will fetch higher economic returns to often costly investments in biotechnology research and development.

Commodities such as bananas plantations, cocoa, oil palm, roots and tubers, which are of high socio-economic importance to Africa, should be on priority biotechnology research agendas of countries in the region. International agencies and research centres should also pay due attention to "orphan" commodities.

Most of the African countries with an interest in modern biotechnology are aware of the need to institute adequate biosafety provisions and establish appropriate intellectual property and patenting systems.

The GATT agreements have brought these issues to the freefront, and they must be attended to in time to enable countries to access global markets. Preliminary work on developing biosafety guidelines is underway in many African countries.

The status of the issue of intellectual property rights in the region is still in a state of flux. Policymakers and biotechnologists in the region, in consultation with concerned international agencies, need to address this issue more decisively. Access to information by establishing the necessary national and regional databases and information systems should be strengthened to render priority setting, technology assessment and transfer mechanisms more effective.

Bacteriological Changes

The Bacterial Flora on Live Fish

Microorganisms are found on all the outer surfaces (skin and gills) and in the intestines of live and newly caught fish. The total number of organisms vary enormously and Liston (1980) states a normal range of 10^2–10^7 cfu (colony forming units)/cm^2 on the skin surface. The gills and the intestines both contain between 10^3 and 10^9 cfu/g .

The bacterial flora on newly-caught fish depends on the environment in which it is caught rather than on the fish species . Fish caught in very cold, clean waters carry the lower numbers whereas fish caught in warm waters have slightly higher counts. Very high numbers, *i.e.*, 10^7 cfu/cm^2 are found on fish from polluted warm waters. Many different bacterial species can be found on the fish surfaces. The bacteria on temperate water fish are all classified according to their growth temperature range as either psychrotrophs or psychrophiles. Psychrotrophs (cold-tolerant) are bacteria capable of growth at 0°C but with optimum around 25°C.

Psychrophiles (cold-loving) are bacteria with maximum growth temperature around 20°C and optimum temperature at 15°C . In warmer waters, higher numbers of mesophiles can be isolated.

The microflora on temperate water fish is dominated by psychrotrophic Gram-negative rodshaped bacteria belonging to the genera *Pseudomonas, Moraxella, Acinetobacter, Shewanella* and *Flavobacterium.*

Members of the *Vibrionaceae (Vibrio* and *Photobacterium)* and the *Aeromonadaceae (Aeromonas* spp.) are also common aquatic bacteria and typical of the fish flora. Gram-positive organisms as *Bacillus, Micrococcus, Clostridium, Lactobacillus* and coryneforms can also be found in varying proportions, but in general, Gram-negative bacteria dominate the microflora.

Shewan (1977) concluded that Gram-positive *Bacillus* and Micrococcus dominate on fish from tropical waters. However, this conclusion has later been challenged by several studies which have found that the microflora on tropical fish species is very similar to the flora on temperate species. A microflora consisting of *Pseudomonas, Acinetobacter, Moraxella* and *Vibrio* has been found on newly-caught fish in several Indian studies . Several authors conclude, as Liston (1980), that the microflora on tropical fish often carry a slightly higher load of Gram-positives and enteric bacteria but otherwise is similar to the flora on temperate-water fish.

Aeromonas spp. are typical of freshwater fish, whereas a number of bacteria require sodium for growth and are thus typical of marine waters. These include *Vibrio, Photobacterium* and *Shewanella.* However, although *Shewanella putrefaciens* is characterized as sodium-requiring, strains of *S. putrefaciens* can also be isolated from freshwater environments. Although *S. putrefaciens* has been isolated from tropical freshwaters, it is not important in the spoilage of freshwater fish.

Table: Bacterial flora on fish caught in clean, unpolluted waters

Gram-negative	Gram-positive	Comments
Pseudomonas	*Bacillus*	
Moraxella	*Clostridium*	
Acinetobacter	*Micrococcus*	
Shewanella putrefaciens	*Lactobacillus*	
Flavobacterium	Coryneforms	
Cytophaga		
Vibrio Photobacterium		*Vibrio* and *Photobacterium*
Aeromonas		are typical of marine waters; *Aeromonas* is typical of freshwater

In polluted waters, high numbers of *Enterobacteriaceae* may be found. In clean temperate waters, these organisms disappear rapidly, but it has been shown that *Escherichia coli* and *Salmonella* can survive for very long periods in tropical waters and once introduced may almost become indigenous to the environment . The taxonomy of S. *putrefaciens* has been rather confused. The organism was originally associated with the *Achromobacter* group but was later placed in the Shewan *Pseudomonas* group IV.

Based on percentage of guanine+ cytosine (GC%) it was transferred to the genus *Alteromonas,* but on the basis of 5SRNA homology it was reclassified to a new genus, *Shewanella.* It has recently been suggested that the genus *Aeromonas* spp. which was a member of the *Vibrionaceae* family be transferred to its own family, the *Aeromonadaceae* . Japanese studies have shown very high numbers of microorganisms in the gastrointestinal tract of fish, and as such numbers are much higher than in the surrounding water, this indicates the presence of a favourable ecological niche for the microorganisms. Similarly, Larsen *et al.* (1978) reported up to 10^7 cfu/g of vibrio-like organisms in the intestinal tract of cod and Westerdahl *et al.* (1991) also isolated high numbers of vibrio-like organisms from the intestines of turbot. *Photobacterium phosphoreum* which can be isolated from the surface can also be isolated in high numbers from the intestinal tract of some fish species. On the contrary, some authors believe that the microflora of the gastrointestinal tract is merely a reflection of the environment and the food intake.

Microbial Invasion

The flesh of healthy live or newly-caught fish is sterile as the immune system of the fish prevents the bacteria from growing in the flesh. When the fish dies, the immune system collapses and bacteria are allowed to proliferate freely. On the skin surface, the bacteria to a large extent colonize the scale pockets. During storage, they invade the flesh by moving between the muscle fibres. Murray and Shewan (1979) found that only a very limited number of bacteria invaded the flesh during iced storage.

Ruskol and Bendsen (1992) showed that bacteria can be detected by microscope in the flesh when the number of organisms on the skin surface increases above 10^6 cfu/cm^2. This was seen at both iced and ambient temperatures. No difference was found in the invasive patterns of specific spoilage bacteria *(e.g.,* S. *putrefaciens)* and non-spoilage bacteria.

Since only a limited number of organisms actually invade the flesh and microbial growth mainly takes place at the surface, spoilage is probably to a large extent a consequence of bacterial enzymes diffusing into the flesh and nutrients diffusing to the outside.

Fish spoil at very different rates, and differences in surface properties of fish have been proposed to explain this. Skins of fish have very different textures. Thus whiting *(Merlangius merlangus)* and cod *(Gadus morhua)* which have a very fragile integument spoil rapidly compared to several flatfish such as plaice that has a very robust dermis and epidermis. Furthermore, the latter group has a very thick slime layer, which includes several antibacterial components, such as antibodies, complement and bacteriolytic enzymes.

4

Genetics in Fisheries

Introduction

Genetically modified fish pose economic benefits to aquaculture, but also pose many ecological risks. Since most of the risks are unknown, further research, review, and monitoring is important. Conducting thorough studies under near natural environmental conditions provides our only hope of assessing the risks associated with the introduction of genetically modified fish. Demand to produce transgenic fish is only increasing, so to improve the chances that decisions to use transgenic fish outside the laboratory are made in an ecologically sound manner, ecological assessments and studies need to be initiated at this time. Minimization of the risks associated with genetically modified organisms can be addressed through the processes of risk assessment and risk management, and by the acceptance of precautionary principles. If we do not approach this subject with caution, the introduction of genetically modified fish could have detrimental affects that are uncontrollable, permanent and irreversible.

In the field of transgenics, there have been numerous methods employed in the transfer of foreign DNA to target organisms, and each has had varying success depending upon the recipient animal. To date, microinjection is still the preferred method for aquatic species, and many of the techniques are modified versions of those used in mammalian microinjection. However, significant differences in fish egg physiology (*i.e.*, a large yolk, an impenetrable outer membrane, and overall opaqueness) make it impossible to inject DNA directly into the nucleus, which renders this method time consuming, labour intensive, and technically challenging.

Electroporation is an alternative that alleviates some of these difficulties and could greatly facilitate research and commercial development of transgenic animals. However, techniques for electroporation have not been standardized, and considerable variation exists among methods. Initial studies involving electroporation in aquatic species used a direct current (DC) with either a pulse of exponential decay or multiple rectangular pulses; however, levels of transfer were low, and expression questionable. As a modification to cell culture electrotransfection techniques, Chang combined the properties of a DC field and an oscillating field to produce a new waveform, called DC-shifted radio frequency (RF) pulses. The idea is that this new waveform would be gentler to the membrane and therefore increase the probability of the membrane resealing.

We recently tested the efficacy of DC-shifted RF pulses to improve gene transfer in fish embryos and to determine if transgene expression could be achieved. The Japanese medaka (Oryzias latipes) was chosen as the study organism. Medaka offer several advantages as model organisms. For example, the medaka is a small, egg-laying freshwater teleost with a rapid generation time, large number of offspring, and external fertilization. Many aspects of this species have been well characterized, including physiology, genetics, and development. Medaka are easy to maintain in a laboratory setting, fertilization time can be predicted, and the fertilized embryos develop externally.

We used the commercially available Gene Pulser II and RF module (Bio-Rad Laboratories, Hercules, CA) with the reporter gene constructs pCMV-SPORT-ß-gal (Invitrogen, Carlsbad, CA) and the pEGFP-1 promoterless plasmid linked to a gene of agricultural importance. The pCMV-SPORT-ß-gal plasmid contains the ß-galactosidase gene from E. coli driven by the CMV promoter and followed by the SV40 polyadenylation signal. The EGFP construct consisted of the CMV promoter driving expression of both the egfp gene and the phyA phytase gene from Aspergillus niger. Plasmid DNA was purified using a commercial kit. Following fertilization, embryos were collected as quickly as possible. Embryos were rinsed with deionized water, separated into groups, and placed into a 0.4cm gap-width cuvette with supercoiled plasmid DNA resuspended in buffer.

Over 15,000 embryos were electroporated with pCMV-SPORT-ß-gal in order to determine how each parameter (buffer choice, buffer volume, quantity of DNA, total voltage, percent modulation of the wave, the radio frequency of the wave, number of bursts, length of bursts, time between bursts, and number of embryos) affected gene transfer. Following electroporation, embryos were rinsed with deionized water, placed within a homemade hatchery, and allowed to develop. At two weeks of age, fry were sacrificed, fixed, and stained overnight to detect ß-galactosidase activity.

Transgenic fry were compared to wild type controls, and blue tissue staining was used as an indicator of transgene expression. Total voltage significantly affected survival of electroporated embryos, and this effect was influenced by buffer choice and volume. The radio frequency significantly affected the number of transgenics obtained, and this effect was most significant with 100% modulation of the wave. The length of the bursts as well as the amount of time between bursts also significantly affected the number of transgenic fry obtained.

Based upon these results, we concentrated on the effects of voltage, radio frequency, burst duration, and burst interval on gene transfer in embryos electroporated in groups of 50 with 4mg of plasmid DNA resuspended in 400ml of HEPES buffered saline solution. From these results, it is evident that voltage will not only affect survival, but in turn will also affect the probability of surviving fry being transgenic. For the most part, survival decreased as voltage increased, yet the percentage of fish with transgene expression increased as voltage increased up to 20-25V and then decreased. Similarly, as burst duration increased, survival decreased.

However, the effect of burst duration on the production of transgenic fish was dependent upon the radio frequency. When the radio frequency was between 35 and 45KHz, the highest percentage of transgenics was obtained with a burst duration of 10msec; for a radio frequency of 30 or 50KHz, the highest percentage of transgenics occurred with a burst duration of 50msec. In contrast, the effect of the burst interval on the production of transgenic fish was linear and did not appear to be influenced by any other factor. As the burst interval increased, the efficiency of gene transfer decreased.

Since the expression patterns in the ß-galactosidase stained fish displayed a high level of mosaicism, we decided to test for germ-line integration, transmission, and expression of the transgene in subsequent generations.

The EGFP construct was utilized because its expression can be observed in living organisms with minimal discomfort to the fish. Following analysis of the above parameters and their effects, we decided to use the same DNA concentration resuspended as above, 25V, 100% wave modulation with a radio frequency of 35KHz, and 3 bursts for 10msec each with a 1.0sec pause between bursts. These fish were grown to sexual maturity and a double back-cross performed. Approximately 85% of the founders that survived to sexual maturity demonstrated germ-line transmission to their offspring3. The observance of green fluorescent protein indicated transgene expression, and Southern blots demonstrated heritability of both the copy number and junction fragments of the transgene in a Mendelian fashion.

It is evident that electroporation has the potential to increase greatly the efficacy of transgenic fish production. We were able to use an electroporation device based upon the principle of a direct current with an oscillating field, called DC-shifted RF pulses, to produce effectively and efficiently germ-line transgenic medaka. Unlike microinjection, this technique is very rapid and easy to use, allowing for as many embryos as can be collected to be electro-porated in a day. Although some time is required to optimize the electroporation condition, once a set of conditions has been determined, it can be used repeatedly to produce transgenic fish with varying transgenes. The high percentage of transgenic fish obtained should allow for an abundant number of founders to be screened for proper or ideal expression, as well as to decrease the number of embryos needing to be sacrificed in order to obtain one transgenic one. Unfortunately, the RF module from Bio-Rad Laboratories is no longer commercially available.

Comparative of fish development

Comparative of fish development constitute one of the glories of nineteenth century embryology. The comparative method as a formal concept in morphology has its roots in German Naturphilosophie and was extended to embryology by Dollinger and von Baer. Subsequently, Johannes Muller and others incorporated it into physiology. Already well established, comparative embryology was thrust into a position of prominence with the publication of Darwin's On the Origin of Species. Darwin stated that much could be learned about evolution through embryological studies.

Fritz Mller and Haeckel considered that embryology, especially comparative embryology, was paramount in understanding evolution, *i.e.*, "ontogeny recapitulates phylogeny". The new evolutionary impetus combined with the pre-existing research agenda drove comparative studies of fish development ahead at a rapid clip. Fishes provided a intellectual feast for evolutionary comparative embryologists. The four major groups of fishes occupy key positions in vertebrate evolution. Agnaths (hagfish, lampreys) are the most primitive representatives of the craniate-vertebrate lineage. Chondrichthyans (sharks, rays, chimaeras) are among the oldest extant jawed vertebrates and incorporate the definitive vertebrate Bauplan and developmental program. Sarcopterygians comprise lungfishes, coelacanths, and tetrapods. The Actinopterygian or ray-finned fishes (Polypterus, sturgeons, gars, Amia, and teleosts) represent a group of bony fishes that evolved in a lineage that culminated in the teleosts, one of the most successful groups of vertebrates. Not only was there evolution at the phyletic level, but the process of development itself underwent evolutionary change.

Comparative venture

Once embarked upon the comparative venture, the availability of study material became a problem. Many fishes of interest were found only in the fresh waters of the tropics, and some, such as the Australian lungfish

Neoceratodus, were only discovered in the last quarter of the nineteenth century. Coincident with the rise of interest in comparative embryology, geographical regions that harboured fish species of interest became more accessible, the fishes themselves became better known, and the technology of travel improved.

Even so, studying the embryology of fishes under primitive field conditions was a major task. Individuals such as Budgett, Kerr, and Semon are part of the golden age of exploration. The hagfishes and lampreys are the most primitive representatives of the craniatevertebrate lineage.

Both are jawless. Hagfishes are common, marine, bottom-dwelling fish in cold, inshore waters and in deeper waters off continental shelves. Recently, hagfishes have been considered to be the sister group to all remaining vertebrates. Virtually everything that is known about their embryonic development is derived from collections of Eptatretus (=Bdellostoma) stout eggs obtained by Bashford Dean and Franz Doflein in Monterey Bay, California, on the Pacific coast of North America in the 1890's. For all intents and purposes, there have been no modern embryological studies of hagfish except for Gorbman, Gorbman and Tamarin, and Wicht and Northcutt, who used Dean's material. For this discussion, research on the Atlantic hagfish Myxine will not be considered because there is so little of it. Doflein and Dean (1898, 1899) described embryonic development. Dean's 1899 extraordinary monograph gives a definitive account, sumptuously illustrated with colour lithographs, of hagfish development from early cleavage through hatching. Unfortunately, because the monograph is not readily accessible, this research has been overlooked.

Most of Dean's account is based on the observation of whole mounts. Sections of blastula and gastrula stages are illustrated. The eggs are large ellipsoids about 22 mm in length by 8 mm in width. They are heavily yolked, and there is a germinal disc at the animal pole. Cleavage is meroblastic. The process of embryo formation coincides with the epibolic movement of the blastoderm over the yolk mass. The equivalent of a germ ring is present and the embryo is formed along an anterior-posterior axis as the blastoderm moves over the yolk mass. The only section of a gastrula reveals the absence of bottle cells, although this could be a function of the plane of section.

Both a cellular "periblast" and an underlying yolk syncytium are depicted. The neural tube is formed by the closure of the medullary folds. Much information is given on organogenesis and the later phases of embryonic development. Additional descriptions of early cleavage stages are given by Gudger and Smith. A description of development of the head, brain and associated organs based on the analysis of sectioned material was published by von Kupffer. Lampreys are widely distributed throughout the waters of the northern and southern temperate zones.

Chondrichthyan Development

They were known to zoologists such as Home, Rathke, and Miller who worked on aspects of their reproductive anatomy. It is surprising, then, that embryological research was only undertaken in the mid-nineteenth century and that comprehensive studies only appeared in the latter part of the nineteenth and in the early twentieth centuries. The first substantive description of lamprey development was that of Schultze, who amongst other observations demonstrated the somewhat unequal holoblastic cleavage of the egg. Aspects of gastrulation were described by Balfour, Goette, and Hatta. Both Goette and Hatta demonstrated the presence of bottle cells and involution of surface cells through the blastopore. Other contributions of note are those of Scott, von Kupffer, and Selys-Longchamps. Older literature on lamprey development is listed in Ziegler. Piavis provides a modern account of development and a list of the more recent literature. Two main lines of evolution are recognized within the class Chondrichthyes: the holocephalans and elasmobranchs.

When chondrichthyan development was discussed previously, only the elasmobranchs (sharks, rays, and skates) were considered. Now the holocephalans (chimaeras or ratfishes) will be considered. Chimaeras are oviparous, bottom-dwelling marine fishes usually found in the deep waters off the continental shelf, except in those regions where the inshore waters are cold, e.g., the northwest coast of North America and New Zealand. Because of their deep water habitat, it has been difficult to procure their eggs for study.

The major sources of information on chimaeroid development are Schauinsland's and Dean's monographs. Schauinsland studied Callorhinchus mil, whereas Dean's study was mostly based on the eggs and embryos of Hydrolagus (=Chimaera) colliei collected in Monterey Bay, California. Dean's monograph is remarkably complete. Dean begins with an account of the breeding habitats of Hydrolagus and egg case formation and then describes in detail development from fertilization through hatching. Chimaeras are oviparous.

Cleavage and Subsequent Development

Their eggs, moderately large, oblate spheroids (25 x 20 mm), are enclosed in an elaborate egg case composed of structural proteins that differ considerably from those of elasmobranchs. Cleavage and subsequent development is meroblastic and closely resembles that of sharks. A closely spaced series of stages from early cleavage through the mid-phases of development is illustrated with surface views of the blastoderm and drawings of the embryos. This series is supplemented with many illustrations of sectioned material, especially from the blastula and gastrula stages. One developmental characteristic that sets chimaeras apart from elasmobranchs is the pronounced elongation of the post-anal region of the body, which begins at the tailbud stage and soon gives the young embryo an almost eel-like appearance. The studies of Schauinsland and Dean remained unchallenged for many years.

Only in the late twentieth century has interest in chimaeroid development revived. The osteichthyans or bony fishes include the sarcopterygians and actinopterygians. The sarcopterygians (fleshy-finned fishes) are now considered to include the tetrapods, two extinct groups of fishes, and two extant groups of fishes: coelacanths and dipnoans (lungfishes). These fishes are of considerable interest because the "lungfishes, coelacanths, or both together are the closest living relatives of tetrapods". The living coelacanth Latimeria chalumnae is the sole surviving representative of a once common group.

The only known population occurs in the Comoros Islands off the east coast of Africa. No embryological research was carried out on Latimeria during the nineteenth century because it was not discovered until 1938. However, in 1975 Smith *et al.* discovered that Latimeria is viviparous when they found five nearterm pups in the oviduct. Up to this time only two gravid females have been collected. Wourms *et al.* described aspects of viviparity and the maternal-embryonic relationship using the embryos and maternal tissues from the Smith *et al.* specimen.

This material was also used by Bemis and Northcutt in their study of the innervation of cranial muscles and Northcutt and Bemis in their study of cranial nerves. Fricke and Frahm described full-term pups obtained from the second gravid female. Coelacanth eggs are enormous, up to 9 cm in diameter and with a dry weight of 184 g. They are heavily yolked and contain yolk platelets. Although no information is available on early embryonic development, it is reasonable to assume that development is meroblastic. Because the coelacanth is on the verge of extinction, it is highly unlikely that any substantial body of embryological knowledge will be forthcoming. The dipnoans or lungfishes comprise the other extant group of sarcopterygian fishes.

Teleostean Embryology

The history of teleostean embryology. The main emphasis of early studies was placed on defining the meroblastic pattern of teleostean development and piscine features of embryogenesis and organ and tissue formation. Some attention was given to the diverse embryonic adaptations for viviparity in different taxa. By the end of the nineteenth century, it was apparent that the early development of most teleosts was very similar but that mid-and late-phases diverged, yielding diverse embryonic morphology and larval forms adapted to specific life styles and habitats.

Thus, the later phases of embryonic development, larval development, and life histories became the subjects of comparative studies of teleostean development during the late nineteenth century and the twentieth century. Several factors favoured the growth of a comparative approach, especially the advent and rise of marine biology, biological oceanography, and fisheries biology. Fishes are the dominant group of marine vertebrates and are commercially important. Thus, the study of life histories and the embryonic and larval stages of inshore, pelagic, and deep-sea fishes became a research agenda for the marine sciences.

The results were often quite spectacular, e.g., the leptocephalus larvae of eels and the bizarre larvae of the stomiatoid fish Idiacanthus with its stalked eyes and trailing gut. Studies of this type were pioneered by Alexander Agassiz and Ryder in the United States, McIntosh and Prince in Great Britain, and Ehrenbaum in Germany, amongst others. Similar activities were undertaken at limnological stations and in freshwater fisheries biology. Popularization of the aquarium hobby and the successful reproduction in captivity of freshwater fishes imported from Africa, South America, and Asia such as the zebrafish, Danio rerio, established another source of material for comparative studies. Given their numbers and diversity, a wealth of information, mostly of a descriptive nature, was produced on the development of a great many species of teleosts. Increased activity in the study of teleostean development was accompanied by the elaboration of several concepts and theoretical approaches which both aided and impeded further progress.

In 1877, Haeckel set forth his Gastrea theory. Based on his observations of teleostean development, he applied the concept to both teleosts and chondrichthyans. Much of the literature on fish development from his time to the present has been concerned with the morphogenetic movements and embryo forming mechanisms that are associated with gastrulation, a literature too voluminous to review. One major problem that arose from overzealous and inept efforts to use embryological studies to support evolutionary theory was the attempt to homologize the events and processes of gastrulation in teleosts and chondrichthyans with amphibians and amniotes.

In this respect, it was Goette who first proposed that involution, *i.e.*, flowing of a sheet of cells over an inpocketing such as a blastopore, has a prominent role in teleost gastrulation. At the risk of oversimplification, many of these studies were comparative in name only. There was little or no conceptual basis for interpreting the mass of information. Attempts, such as that of Schultz, to establish a well-defined conceptual basis for comparative embryological research were overlooked. (In passing, Schultz's 1910 monograph appears to be a direct precursor of de Beer's 1930 study of embryology and evolution.) Although the body of evolutionary theory was available, the influence of Haeckel's biogenetic law interfered with the productive application of it to many developmental issues.

A growing anti-evolutionary bias tended to discourage this type of inquiry. Studies of individual species were often exceedingly well done, and the development of a single organ system was often compared in several different taxa and interpreted in the context of the evolutionary paradigm.

Examples of the latter include Ryder's study of heterocercy and the evolution of the fins and fin rays of fishes and Eigenmann's study of the abortive development of the eye in blind cave fishes. Eigenmann compared the development and degeneration of the eye in the blind cave fish Amblyopsis with that of the surfperch Cymatograster, a fish with a well-developed eye. In Amblyopsis, development of the eye begins normally, but then the lens degenerates and the entire eye regresses. Recently, Jeffery, using the embryos of another blind cave fish Astyanax, has shown that programmed cell death of the lens eliminates lens induction from the cascade of inductive interactions required for normal eye development.

In general, however, the comparative method was not used to explore the evolution of development and the factors that brought about evolutionary change. It was not until the maturation of evolutionary thought in the twentieth century that a new set of concepts and approaches were produced and used to exploit the potential of comparative studies. Paramount among these are: (1) revived interest in the evolution of development and the role of development in evolution; (2) application of evolutionary theory to life histories; and (3) attempts to order the bewildering diversity of fish reproduction in terms of systematics or reproductive guilds.

Method of artificial fertilization

Rosconi gives a detailed account of his method of artificial fertilization. Developing eggs, thus obtained, were of known age and were subsequently examined microscopically. His major contribution is that he appears to have been the first to describe cleavage in the teleostean egg. His detailed description is illustrated by a plate which "rivals all other in the literature for accuracy and beauty in presenting the early stages of teleostean development". He also made extremely accurate observations on the embryonic shield and its role in the formation of the embryo. To aid observation, he used an acid to stop development and render the embryo opaque without affecting the transparency of the yolk. His figures depict the concentration of cells in the midline of the embryonic shield that are the precursor of notochord and spinal cord. Elsewhere, he states for the first time that the teleostean nerve cord originates from a single structure rather than by the folding of a neural plate.

Embryological research

Embryological research on teleosts moved to France during the middle years of the century. The first important work was Vogt's contribution to the embryology of the whitefish (Coregonus), which was part of Agassiz's elaborate series of monographs on the freshwater fishes of central Europe.

The embryology portion consists of a text volume and a folio of coloured lithograph plates. Although considered part of the French School by Oppenheimer, it could just as well be considered a product of the Dollinger-Wurzburg School. Agassiz was in close contact with Dollinger at Munich between 1827 and 1830 and was studying teleost development.

The role of Agassiz, if any, in Vogt's study is not clear, for the dispute over authorship was the first in a series of controversies that would occur between Agassiz and his collaborators. In addition to its scientific methods (vide infra), this work had an extraordinary impact because of the clarity and beauty of its illustrations. The figures were drawn by professional artists and reproduced, many in colour, by lithography, a printing process capable of rendering fine detail. Not only were new observations effectively presented, but also the known details of development took on a new lucidity.

The Vogt-Agassiz study is another study in which developing eggs were obtained by artificial fertilization, cultured within a set temperature range (5-IOC), and examined microscopically. For the first time, the technique of dechorionation is mentioned as an aid for observation. Vogt also used the technique of acid fixation to enhance the visibility of the embryo. A series of well-illustrated, closely spaced stages in the normal development of the whitefish was presented.

This work appears to be the first study to apply a cellular approach to the interpretation of developmental events, especially oogenesis, cleavage, epiboly, and embryogenesis. Information is provided on oogenesis, the structure of the mature egg, and fertilization. According to Oppenheimer, Vogt was the first to inquire into the mechanics of embryo formation. He also speculated on the role of cell movement and cell migration in development. A considerable portion of the work was devoted to the microanatomy of the later stages of embryonic and larval development. In this respect, Vogt gives an account of the formation of a number of organ systems, tracing their development from the earliest cellular condensations or anlagen to completion.

Illustrations of melanocytes and the cellular components of the notochord and connective tissue are remarkable for their detail which included nuclei. The next figure of consequence in the French school was Lereboullet, a pupil of Duvernoy, who in turn was a protege of Cuvier. Lereboullet made substantial contributions in the field of descriptive embryology and subsequently in experimental studies which were the first great attempt to analyse the mechanism of embryo formation.

His monograph on the development of the pike and perch established new standards in terms of the accuracy with which his observations were depicted. He provided the first detailed description of the microscopic organization of the egg and a more accurate description and illustration of the embryonic shield.

He stated that it is the precursor of the entire embryo and not just the central nervous system. His most important single observation was the description of the yolk syncytial layer (periblast) which he termed "feuillet muqueux," literally mucous lamina. He mistakenly ascribed to it the fate of forming elements of the digestive system. He regarded the germ ring, *i.e.*, the thickened margin of the blastoderm, as a dynamic force during the formation of the embryo. This viewpoint, reinforced by his studies on experimental teratology (vide infra), led him to interpret the process of embryo formation in terms of concrescence, *i.e.*, the right and left portions of the germ ring assemble into the two halves of the body. Lereboullet arrived at this conclusion a decade before His formally promulgated the concrescence theory.

Comparative embryology

Finally, in his monograph on the comparative embryology of the trout, a lizard and a snail, he refuted the concept of a unity in the plan of development propounded by Meckel and Serres. Three important structures, characteristic of teleostean development, still need to be dealt with, namely the yolk syncytial layer or periblast, the enveloping layer, and Kupffer's vesicle. Although Lereboullet first described the yolk syncytial layer, it was von Kupffer who suggested that the cells of the blastoderm not only give rise to the tissue cells of the embryo but also are progenitors of the yolk syncytial layer. Agassiz and Whitman (1884, 1915) proved beyond any doubt that yolk syncytial nuclei are derived from the marginal cells of the blastoderm.

The enveloping cell layer is a tightly sealed epithelial monolayer that forms the outermost boundary of the blastoderm. Priority for its discovery is unclear, for descriptions of it under various names appeared more or less simultaneously in a number of independent studies. Kupffer's vesicle is a transient structure that is only found in teleost embryos. It was discovered by von Kupffer, who thought that it might be an allantoic rudiment. It is a small, fluid-filled epithelial sac that appears during the somite stages. Kupffer's vesicle develops mid-ventrally in a region below the tail bud and at the interface between the embryo and yolk mass. Even now, its function, if any, still remains unknown. At this point, a digression to consider the development of viviparous teleosts is in order. Following the pioneering studies of Forchhammer and Rathke on Zoarces, knowledge of viviparity in teleosts progressed steadily through the rest of the century.

In 1844, Duvernoy described the development of a freshwater poeciliid fish from Surinam. His work was mainly concerned with the later stages of development, but did provide some important information on early development.

The ovary is the site of gestation. Fertilization takes place in the follicle where the embryo is invested by an egg envelope. Development to term takes place in the follicle. Duvernoy illustrated for the first time an early stage follicular embryo in its egg envelope. In his Memorias, published 1851-55, the Cuban naturalist Poey was able to extend the number of viviparous species to include all ten species of Cuban poeciliids. Of greater significance were his observations for the first time of sperm storage in these fishes.

Unfortunately, he interpreted his observations in terms of parthenogenesis rather than sperm storage and superfetation. The problem was not satisfactorily resolved until 1908, when Phillipi discovered that in poeciliids, sperm is packaged into spermatozeugmata that are transferred to the male reproductive tract. He also presented evidence for specific structural pathways, which he called "delle," that allow sperm to move from the lumen into close proximity with the egg and fertilize it within the follicle. In 1856, Wyman produced a definitive study on the embryology of the foureyed fish Anableps. It, too, is a fish with follicular gestation.

He depicts the elaborate follicular epithelium and the highly vascularized abdominal trophoderm of the embryo. Because of the massive increase in size of the embryo during gestation, he postulated that the developing embryos are nourished by secreted maternal proteins. He extended his speculations to include embryonic nutrition in the surfperch and the electric ray Torpedo. In 1853, Agassiz first described viviparity in the surfperches or embiotocids from the Pacific coast of North America.

In a more extensive publication, Girard described and illustrated 17 species. He provided the first detailed information about their reproduction and development. Gestation is ovarian: embryos develop in the ovarian lumen and are enfolded by sheaths of ovarian tissue. Initial egg size is minute, but term embryos are large. A series of developmental stages is described and illustrated. Figures on brood size which varies (10-80 young) according to species are given. The duration of gestation is estimated at 5-6 months. The initial descriptions of Agassiz and Girard culminated in a definitive study of surfperch development by Eigenmann which is tour de force of descriptive embryology.

He noted that the early cleavage stages of the small, yolkpoor eggs verge on being holoblastic. He described the presence of sperm in the ovarian fluid and the imbibition of ovarian fluid by the embryos. Male embryos undergo precocious sexual development. In an earlier separate publication, he described the precocious segregation of germ cell determinants at an early cleavage stage, an overlooked, early contribution to concepts of germ cell development. Throughout the remaining years of the century the occurrence of viviparity was described in various groups of teleost, e.g., the rockfish, Sebastes, halfbeaks, Hemirhamphus, and the Lake Baikal oilfish, Comephorus.

One should recall that such studies are still continuing, with viviparity in the living coelacanth Latimeria not discovered until 1975. By 1886, the field had progressed to the extent that Ryder could review it and add to it some of his own original observations. He concluded that the developing embryos of surf perches derive nutrients by absorbing imbibed ovarian fluid through elongated intestinal villi found in the hypertrophied hindgut. He also concluded that the other characteristic feature of surfperch embryos, the unique vascular apparatus of the enlarged vertical fins, serves a respiratory rather than a nutritive function. Much of the research on viviparity, both in teleosts and chondrichthyans during the latter part of the century was concerned with the elaboration of highly diverse maternal and embryonic structural adaptations for viviparity and attempts to correlate these structures with physiological function, e.g., nutrition, gas exchange.

Advances in technology had profound effects on embryological research during the latter half of the century. Because of the transparency of most teleostean eggs observations had usually been made on intact eggs or microdissections of embryos. Beginning in 1860, a series of discoveries and innovations in histological microtechnique made it possible to fix, embed, section, stain, and subsequently examine serial sections of embryonic material with microscopes equipped with high resolution, apochromatic, oil-immersion objectives (cf. Ryder, 1884b and Whitman, 1885 for the availability of techniques in embryology, and Horder *et al.*, 1985, for a chronology of technological and conceptual innovations). In this respect, the application of photography to the study of fish development occurred surprisingly early in the history of photography. Von Kupffer is credited with the first extensive use of photomicrographs to illustrate an embryological monograph in his 1878 study of herring development. By 1898, His had published micrographs of sectioned trout blastomeres that illustrated aspects of chromosome behaviour during cell division.

The consequences of these advances in embryological technology was to trigger a veritable explosion of embryological studies of teleosts based on the microscopic study of sectioned material. By the end of the nineteenth century, an excellent descriptive knowledge of teleost development had been achieved. But there was a trade-off.

The emphasis on the study of sectioned material detracted from the observation of living embryos, especially cell behaviour during early development, a field that was just beginning to be explored by His (vide infra). This trend led to subsequent difficulties in the interpretation of morphogenetic movements and embryo formation.

Even as this trend was established, its advantages and limitations were becoming obvious. Harrison published his histological study of the differentiation and axonal growth of spinal neurons in trout embryos in which he focused on the patterns of growth of individual nerve cells and their role in establishing neural connections. The constraints of the histological approach led him to develop tissue culture and use it six years later to investigate neuronal outgrowth.

Biochemical genetic markers

Biochemical genetic markers have been increasingly used for inferences on the population genetic structure of various fish, but little attention has been paid to relative differences in the distribution of variability within species in different groups of fish. Worldwide, fisheries provide a source of employment for millions of people and a source of protein for billions. The management of fisheries is therefore a critical activity.

Unfortunately, the realisation that fish are a limited resource was slow to reach acceptance, and there is not much evidence that attempts to manage fisheries have led to sustainability. In 1865, a group led by T.H. Huxley published the Sea Fisheries Commission Report into British fisheries, the first of its kind, in which it was recommended that there should be as few restrictions as possible on inshore and offshore fisheries and the fishing gear used. In effect, the conclusion of the report was that the supply of fish was inexhaustible.

One hundred years later it was clear that this was patently not the case, and in more recent times many of the world's major fisheries have collapsed – or are on the verge of doing so – through overfishing and poor management.

In order to manage a fishery effectively, a great deal of information is required. We need to know, for example, the size of a population, its habitat and migratory behaviour, its age and size structure, the reproductive pattern of the species, the natural mortality rate, the rate at which fish are removed by fishing, and so on. Most of these factors can be, and are, determined from fishery statistics based on detailed surveys and analyses of landings. However, a critical requirement is to know whether the fish species exists as a single genetic unit or as a series of relatively genetically distinct groups. Is the species genetically homogeneous (panmictic) or is it genetically heterogeneous? If there are local genetically distinct groups, management strategies will need to be adapted to take this into account.

genetically discrete groups

This substructuring into different, relatively genetically discrete groups may take various forms. For example, the migratory salmonids return to the rivers in which their parents spawned and this effectively creates genetic groups that are associated with particular rivers or even particular regions of a river. In another example, some lake fish have autumn and spring spawners, and as long as this character is inherited, two genetically distinct groups can evolve.

In many marine fish, there are particular sites where spawning aggregations occur and these can lead to genetic differentiation. In the case of sedentary shellfish, dispersal is usually mainly via a larval phase, and hydrographic factors will constrain the direction and distance of dispersal. Shellfish such as flat oysters with a relatively short larval phase or other species with direct development will be more likely to be genetically differentiated than shellfish such as cupped oysters or mussels with an extended larval phase. What should we call these genetically differentiated groups and what qualifies as 'genetically differentiated'?

It is possible to quantify levels of genetic differences using various indices, but for the moment, let us consider the terminology that is available. Partly or strongly genetically differentiated groups within a species have been called (1) 'varieties', 'races', 'breeds' or 'strains' if they are cultivated and (2) 'stocks', 'demes', 'populations', 'subpopulations', 'types' or 'ecotypes' if they are wild. All of these terms have been used in the past to describe genetic differentiation within species, but they have often not been clearly defined.

In the case of fisheries, the word 'stock' has often been defined by fishery managers as a group of fish exploited by a specific method or existing in a particular area. Although this may be convenient for the analysis of landings or catch and effort data, and for the enforcement of management measures such as quotas or net size, it may bear little relation to the true genetic substructuring of the fish species. The important point to realise is that there is a continuum of genetic differentiation.

In some species there may be clearly genetically distinct groups, while in others (or in other parts of their range) there will be groups which are virtually indistinguishable genetically. Thus, the use of a single word such as 'stock' to define the units of such substructuring is not often appropriate. Here, we will use the less prescriptive term of 'population' to define an intraspecific (within a species) group of randomly mating individuals which exist (and can therefore be sampled) in a defined geographic position or at a defined time.

A key point to note here is that we are assuming that all individuals comprising a population are able to mate randomly with one another, and therefore, there can be no further genetic subdivision within a population. The main types of genetic markers were outlined, and we now need to show how these markers can be used to identify any population substructure within species of fish or shellfish. We will start with markers such as allozymes, microsatel-lites and single nucleotide polymorphisms (SNPs) that are co-dominant, that is they allow direct identification of homozygotes and heterozygotes.

How do we handle this information on genotypes in a population? Firstly, we should note that the process of meiosis breaks up existing genotypes and that new genotypes are reassembled when egg and sperm combine at syngamy. Therefore, although there will be the same range of genotypes in an offspring generation as in a parent generation, the two are not directly linked by inheritance. What is linked by inheritance, however, is the presence of an allele and the frequency of that allele in the population. And it is the allele frequency (or gene frequency) that is used to investigate the structure of populations within species.

Frequencies Estimated

When genotypes at a co-dominant locus are scored from a sample of individuals, the frequencies of the alleles at the locus can easily be calculated from these data.

Remember that every diploid individual has pairs of chromosomes, so they must also have pairs of alleles at each locus, that is a homozygote at a locus has two of the same allele and a heterozygote has two different alleles. Although a diploid individual can only exhibit two alleles at a locus, there may be many different alleles at that locus present in the population as a whole. Allele frequencies are the critical starting point for all further analyses of genetics in populations. By convention, the frequencies of alleles at a locus are symbolised using the lower-case letters *p*, *q*, *r*, *s* and the frequency of an allele is given by:

$$2H_o + H_e p = 2N$$

where H_o is the number of homozygotes for that allele, H_e is the number of heterozygotes for the allele and *N* is the number of individuals scored at the locus.

Alleles and Genotypes

Considering a single locus, it is convenient to envisage a population as a 'pool' of alleles. Haploid gametes are produced by each generation, and each of these gametes contains a single allele from this pool. Let us assume that all adults in the population provide equal numbers of gametes and that each gamete has an equal chance of combining with all other gametes. Once we know the frequencies of the alleles in the population, it is possible to calculate the frequencies with which the genotypes will be present in the offspring from that generation. This principle was first quantified in 1908 by both Godfrey Hardy, a British mathematician, and Wilhelm Weinberg, a German doctor, and is called, unsurprisingly, the Hardy–Weinberg model.

The random mixing of eggs and sperm the proportions of the genotypes will be:

AA AB BB

$$p^2 \; 2pq \; q^2$$

which, in mathematical terms, is equal to $(p + q)^2$. As long as nothing upsets the Hardy–Weinberg model, allele frequencies should remain constant from generation to generation. The frequencies of the genotypes predicted by the model are dependent on the assumptions of equal and random parental contributions and equal and random mixing of gametes. However, if this is not the case, the model will not hold true. Also, we do not usually sample animals at the beginning of their lives; normally fish or shellfish are sampled as adults. In order for the Hardy–Weinberg model to be appropriate in populations of adults, further assumptions need to be made. We must assume that there is no selection that causes one or other genotype to suffer differential mortality.

We must also assume that our sample taken does not include individuals that have migrated into the population from some other population with different genotype frequencies. So both migration and selection are factors that can cause deviation from the genotype frequencies predicted by the Hardy–Weinberg model. If the number of parents that reproduce in a population is significantly smaller than the whole adult population, allele frequencies in these parents might differ from the whole population and the allele frequencies in the following generation will therefore change. A further consideration is that very few fish or shellfish species exist as discrete generations, most populations being mixtures of animals from different overlapping generations.

Allele Frequencies Change

If a population is in agreement with the Hardy–Weinberg model, then allele frequencies at a locus are not being driven (by selection or migration) to change significantly from generation to generation. However, there is always bound to be some variation in allele frequency from one generation to the next because of the element of chance. Contrary to the idealised situation required to make mathematical predictions, in real life all individuals of all genotypes will not produce exactly the same number of gametes. Neither are all gametes likely to undergo truly random mixing. Therefore, there will be variation from the model in every generation. This is a fact of biological variability, called random genetic drift. In order to quantify this natural variation, allele frequencies can be given a variance as indicated in the formula:

$p(1\text{-}p)$ Variance of frequency of allele = $2N_e$

where p is the frequency of the allele and N_e is the effective population size. The concept of the effective population size is an important one. It is the number of individuals in the population that contribute genetically to the next generation.

It will exclude juveniles, individuals too old to reproduce and those that provide non-contributory gametes. These may be individuals that have infertile gametes, or those whose gametes never come into contact with the gametes of other members of the population due to geographic position or timing of spawning. Thus it can be seen how the effective population size could, in certain circumstances, be very much smaller than the total number of individuals in the population.

If the effective population size (N_e) in the expression above is a very large number, then the allele frequency variance between successive generations will be very small. Conversely, allele frequency variance will be large when N_e is small.

This means that natural fluctuations in allele frequency between generations will be much greater in small populations. We shall encounter this concept again when dealing with various aspects of genetics in small populations.

Allele frequencies in a population can therefore change over time owing to random genetic drift. However, they may also change owing to pressure of selection (one genotype or one allele survives better than others at a locus) or to patterns of migration or dispersal that may fluctuate in direction or in strength over time.

Population Structure Arise

The differentiation of a species into genetically different populations is a fundamental part of the process of evolution. For a number of physical or biological reasons, the distribution of a species may become fragmented. For example, during the last ice ages, aquatic species in the temperate regions of the northern hemisphere were driven south and many became restricted to local areas in southern habitats. Where species were fragmented into different areas, many of these populations remained isolated for long periods of time in such refugia. Following the retreat of the last glaciation some 10 000–13 000 years ago, aquatic and terrestrial species went through the process of gradual recolonisation of habitat and this has provided plenty of opportunity for further fragmentation.

Once fragmentation has occurred, allele frequencies at most loci will be subject to random genetic drift in the fragmented populations. As we have seen, random genetic drift produces more rapid allele frequency change in small, compared with large, populations. If only a few individuals are involved initially in founding a population, then allele frequencies in the new population may be very different from the source population, and if numbers remain small, further allele frequency changes may be rapid.

In addition to random genetic drift, localised adaptation will occur involving selection at some loci for particular characteristics and these will cause further differences between populations. These loci under selection, and to a lesser extent those linked to them, can show higher patterns of differentiation between populations than neutral markers and are sometimes called 'outliers'. These processes of random genetic drift and adaptation will tend to increase differences between populations, while migration or larval dispersal between populations will tend to reduce it. So the population differentiation we can identify today in a species is essentially the result of the historical interplay between the environment and the forces of dispersal, adaptation and random genetic drift.

Genetic improvement technologies

Genetic improvement technologies cover a range of techniques requiring different levels of expertise and resources. Chromosome-set manipulation (*i.e.* polyploidy induction) is an established technique to increase the number of chromosome-sets (ploidy number) in an organism. Temperature, chemical and pressure shocks applied to fish eggs can be used to produce triploid (3 chromosome-sets) individuals that have desirable culture traits. Sex-reversal and the production of single sex groups of fish is also a simple technology that combines hormone treatment and chromosome-set manipulation. Hybridisation, *i.e.* the mating of genetically different groups from the same species (intra-specific hybridisation) or from different species (inter-specific hybridisation), is a simple technique that is now easy to accomplish due to our increased knowledge of reproductive biology.

It can be used to combine good traits from two different species into one group of fish or to transfer a characteristic of one group to another. A problem is that breeding hybrids with hybrids results in a non-uniform and unpredictable group of fish that is generally not well suited for culture. Therefore, for hybrid production, the parent-lines must be maintained pure. The above genetic improvement techniques are considered short-term strategies, where the gains are seen in one or two generations. Selective breeding is a longer-term strategy where gains are accumulated at each generation of selection.

Molecular markers may now increase the efficiency of selective breeding by facilitating the identification of quantitative trait loci (QTLs), *i.e.* genes that control complex characters such as growth rate and environmental tolerance and, secondly, by making it possible to use molecular markers linked to QTLs to identify desirable individuals or families. Genetic engineering and the production of transgenic organisms is an active area of research and development in aquaculture. This is a medium-long term strategy in that development and testing of stable transgenic lines requires time. The large size and hardy nature of many fish eggs allows them to be manipulated rather easily and facilitates gene transfer by direct injection of a foreign gene or by electroporation, where an electric field assists gene transfer. We will briefly discuss currently available biotechnologies in the context of fishery management, aquaculture and conservation respectively.

Applications of FISH Technology in Microdeletion Syndromes

Syndromes such as Prader-Willi syndrome or Smith-Magenis syndrome were first characterized as chromosome deletion syndromes by high resolution cytogenetic analysis of prometaphase chromosomes. These deletions are small and often difficult to detect, or in the case of Williams syndrome, undetectable on prometaphase chromosomes. The term microdeletion syndromes is commonly applied to these.

Cosmid probes have been developed that recognize unique DNA sequences within the critical region of chromosomal deletion for several of the microdeletion syndromes. These probes, and their commercial availability, has made the laboratory diagnosis of these syndromes more accurate. In some cases, such as Velocardiofacial/DiGeorge syndrome, the prevalence of the deletion 22q11.2 both de novo and familial, in the population had not been appreciated. In many labs, FISH is now the test of choice to document these deletions because it is more accurate, and arguably faster to perform. The tests are performed on metaphase chromosomes. Usually the microdeletion cosmid (locus-specific) and a second cosmid (chromosome-specific) are hybridized together to insure identification of the chromosome and to provide an internal control.

Characterization of Chromosome Structural Abnormalities

Combinations of FISH probes and standard cytogenetics can characterize structural rearrangements and marker chromosomes. Take the example of a patient identified in the newborn period who has multiple congenital anomalies consistent with possible trisomy of the long (q) arm of #10. Standard cytogenetic studies demonstrate a structural abnormality of 10q: it is twice the length of the normal homologue. A chromosome #10 whole chromosome paint hybridizes to the normal #10, and to most, but not all of the abnormal #10. This helps to confirm a duplication of chromosome 10 material, and the G-banding pattern is consistent with 10q.

What about the rest of the chromatin that does not hybridize with the paint? Again, by the banding pattern, there is a suggestion that part of chromosome 18 is present. An alpha satellite probe for #18 centromere hybridizes to the abnormal #10. We have a dicentric! Chromosome 18 paint confirms that there are unique, #18-specific sequences present, and the banding pattern suggests 18p and proximal 18q. This child is in fact trisomic for 10q, but also for part of chromosome 18.

This kind of diagnosis would not have been confirmed prior to the advent of clinical molecular cytogenetics. FISH is also teaching us to expect the unexpected: abnormalities are often more complex than they first appear.

Aneuploidy Detection

The autosomal trisomies 21, 13, and 18, together with sex chromosomal aneuploidies, are responsible for well over 90% of the chromosomal disorders found in human conceptuses. They are easily diagnosed by standard metaphase cytogenetic studies of chorionic villus samples (CVS), amniotic fluid cells, and peripheral or cord blood lymphocytes, or other tissue that can be grown in tissue culture.

Chromosome-specific cosmids or alpha satellite probes can be used in FISH studies to document chromosome gain or loss in cells that are not dividing. Interphase cytogenetics using FISH is a rapidly growing field. One can enumerate the number of fluorescent signals present in interphase nuclei, for instance using an alpha satellite probe for the centromere of chromosome 18, or a unique sequence cosmid for a locus on chromosome 21. Some laboratories offer interphase FISH aneuploidy studies of prenatal samples and for newborns suspected of having an autosomal trisomy.

Currently, these studies are used for screening, not as a stand-alone diagnostic test. It is the opinion of most geneticists that interphase FISH studies must be performed in conjunction with standard cytogenetic testing for clinical diagnosis. The results must be interpreted with caution, and appropriate controls must be developed.

We cannot always extrapolate from signal number to the type of chromosome abnormality: three #18 alpha satellite signals in 80 of 100 nuclei may signify Trisomy 18, but we would see the same pattern with an extra isochromosome 18p. We can say that the signal pattern is consistent with a chromosomal abnormality of #18. In this example, have we documented mosaicism? What about those other 20 cells? Most likely, hybridization was not complete in these cells, and not all cells are informative. We must beware of overinterpretation.

Interphase FISH can be applied to noncultured cells and fixed tissue. An interesting use is the detection of tissue limited mosaicism of an aneuploid cell line, for instance in placental tissue or in buccal mucosa. Again, the technology is driving a new appreciation of the extent of chromosomal mosaicism to be found in humans.

FISH and Cancer Cytogenetics

The study of the acquired chromosomal abnormalities in cancer cells continues to expand. Cancer cells are often difficult to grow in culture and FISH techniques can augment standard cytogenetic testing.

Numeric changes such as trisomy 8 in myeloid disorders can be detected by interphase FISH. Certain translocations, such as the bcr/abl rearrangement in chronic myelogenous leukemia and acute lymphoblastic leukemia can be detected by FISH on both metaphase preparations and interphase nuclei.

FISH studies are being used to look for early relapse and residual disease in nondividing cells.

In allogeneic bone marrow transplant patients who received opposite sex donor cells, the success of engraftment can be monitored by FISH chimerism studies.

Differentially labeled X-and Y-specific probes can be used to detect the proportions of XX to XY cells in bone marrow or peripheral blood nuclei in a dual colour FISH procedure. By combining immunocytochemical detection of cancer cells and FISH techniques, chromosomal abnormalities and cell type can be simultaneously studied.

Gene Mapping

FISH is used to map genes to specific chromosomes and chromosomal regions. The order of genes or gene sequences within a chromosome can be established by labeling the probes with different fluorochromes and detecting their hybridization colour pattern. In a new technique called fiber FISH, chromosome-specific chromatin fibers can be spread on a slide and then hybridized with locus-specific probes to allow fine resolution mapping of DNA sequences.

Comparative Genomic Hybridization

Comparative genomic hybridization (CGH) is a molecular cytogenetic technique that combines FISH and standard cytogenetics to look for genome-wide DNA amplification and deletions in tumor cells or cells of a patient with a possible chromosomal disorder.

Its value is that unlike standard cytogenetics, cell culture is not necessary. Like standard cytogenetics, chromosome gains and losses, and amplification or deletion of chromosome regions can be detected without having to selectively FISH for a specific chromosome or sequence. Its use is currently in the research lab. Issues of sensitivity and reliability are being addressed. However, as with other molecular cytogenetic techniques, one year's experiment is next year's clinical test.

Genetically Discrete Groups

This substructuring into different, relatively genetically discrete groups may take various forms. For example, the migratory salmonids return to the rivers in which their parents spawned and this effectively creates genetic groups that are associated with particular rivers or even particular regions of a river. In another example, some lake fish have autumn and spring spawners, and as long as this character is inherited, two genetically distinct groups can evolve.

In many marine fish, there are particular sites where spawning aggregations occur and these can lead to genetic differentiation. In the case of sedentary shellfish, dispersal is usually mainly via a larval phase, and hydrographic factors will constrain the direction and distance of dispersal. Shellfish such as flat oysters with a relatively short larval phase or other species with direct development will be more likely to be genetically differentiated than shellfish such as cupped oysters or mussels with an extended larval phase. What should we call these genetically differentiated groups and what qualifies as 'genetically differentiated'?

It is possible to quantify levels of genetic differences using various indices, but for the moment, let us consider the terminology that is available. Partly or strongly genetically differentiated groups within a species have been called (1) 'varieties', 'races', 'breeds' or 'strains' if they are cultivated and (2) 'stocks', 'demes', 'populations', 'subpopulations', 'types' or 'ecotypes' if they are wild. All of these terms have been used in the past to describe genetic differentiation within species, but they have often not been clearly defined.

In the case of fisheries, the word 'stock' has often been defined by fishery managers as a group of fish exploited by a specific method or existing in a particular area. Although this may be convenient for the analysis of landings or catch and effort data, and for the enforcement of management measures such as quotas or net size, it may bear little relation to the true genetic substructuring of the fish species. The important point to realise is that there is a continuum of genetic differentiation.

In some species there may be clearly genetically distinct groups, while in others (or in other parts of their range) there will be groups which are virtually indistinguishable genetically. Thus, the use of a single word such as 'stock' to define the units of such substructuring is not often appropriate. Here, we will use the less prescriptive term of 'population' to define an intraspecific (within a species) group of randomly mating individuals which exist (and can therefore be sampled) in a defined geographic position or at a defined time.

A key point to note here is that we are assuming that all individuals comprising a population are able to mate randomly with one another, and therefore, there can be no further genetic subdivision within a population. The main types of genetic markers were outlined, and we now need to show how these markers can be used to identify any population substructure within species of fish or shellfish. We will start with markers such as allozymes, microsatel-lites and single nucleotide polymorphisms (SNPs) that are co-dominant, that is they allow direct identification of homozygotes and heterozygotes.

How do we handle this information on genotypes in a population? Firstly, we should note that the process of meiosis breaks up existing genotypes and that new genotypes are reassembled when egg and sperm combine at syngamy. Therefore, although there will be the same range of genotypes in an offspring generation as in a parent generation, the two are not directly linked by inheritance. What is linked by inheritance, however, is the presence of an allele and the frequency of that allele in the population. And it is the allele frequency (or gene frequency) that is used to investigate the structure of populations within species.

Frequencies Estimated

When genotypes at a co-dominant locus are scored from a sample of individuals, the frequencies of the alleles at the locus can easily be calculated from these data. Remember that every diploid individual has pairs of chromosomes, so they must also have pairs of alleles at each locus, that is a homozygote at a locus has two of the same allele and a heterozygote has two different alleles. Although a diploid individual can only exhibit two alleles at a locus, there may be many different alleles at that locus present in the population as a whole. Allele frequencies are the critical starting point for all further analyses of genetics in populations. By convention, the frequencies of alleles at a locus are symbolised using the lower-case letters *p, q, r, s* and the frequency of an allele is given by:

$$2H_o + H_e p = 2N$$

where H_o is the number of homozygotes for that allele, H_e is the number of heterozygotes for the allele and N is the number of individuals scored at the locus.

Alleles and Genotypes

Considering a single locus, it is convenient to envisage a population as a 'pool' of alleles. Haploid gametes are produced by each generation, and each of these gametes contains a single allele from this pool. Let us assume that all adults in the population provide equal numbers of gametes and that each gamete has an equal chance of combining with all other gametes. Once we know the frequencies of the alleles in the population, it is possible to calculate the frequencies with which the genotypes will be present in the offspring from that generation. This principle was first quantified in 1908 by both Godfrey Hardy, a British mathematician, and Wilhelm Weinberg, a German doctor, and is called, unsurprisingly, the Hardy–Weinberg model.

The random mixing of eggs and sperm the proportions of the genotypes will be:

AA AB BB

$$p^2 \; 2pq \; q^2$$

which, in mathematical terms, is equal to $(p + q)^2$. As long as nothing upsets the Hardy–Weinberg model, allele frequencies should remain constant from generation to generation.

The frequencies of the genotypes predicted by the model are dependent on the assumptions of equal and random parental contributions and equal and random mixing of gametes. However, if this is not the case, the model will not hold true. Also, we do not usually sample animals at the beginning of their lives; normally fish or shellfish are sampled as adults. In order for the Hardy–Weinberg model to be appropriate in populations of adults, further assumptions need to be made. We must assume that there is no selection that causes one or other genotype to suffer differential mortality.

We must also assume that our sample taken does not include individuals that have migrated into the population from some other population with different genotype frequencies. So both migration and selection are factors that can cause deviation from the genotype frequencies predicted by the Hardy–Weinberg model. If the number of parents that reproduce in a population is significantly smaller than the whole adult population, allele frequencies in these parents might differ from the whole population and the allele frequencies in the following generation will therefore change.

A further consideration is that very few fish or shellfish species exist as discrete generations, most populations being mixtures of animals from different overlapping generations.

Allele Frequencies Change

If a population is in agreement with the Hardy–Weinberg model, then allele frequencies at a locus are not being driven (by selection or migration) to change significantly from generation to generation. However, there is always bound to be some variation in allele frequency from one generation to the next because of the element of chance. Contrary to the idealised situation required to make mathematical predictions, in real life all individuals of all genotypes will not produce exactly the same number of gametes. Neither are all gametes likely to undergo truly random mixing. Therefore, there will be variation from the model in every generation. This is a fact of biological variability, called random genetic drift. In order to quantify this natural variation, allele frequencies can be given a variance as indicated in the formula:

$p(1\text{-}p)$ Variance of frequency of allele = $2N_e$

where p is the frequency of the allele and N_e is the effective population size. The concept of the effective population size is an important one. It is the number of individuals in the population that contribute genetically to the next generation.

It will exclude juveniles, individuals too old to reproduce and those that provide non-contributory gametes. These may be individuals that have infertile gametes, or those whose gametes never come into contact with the gametes of other members of the population due to geographic position or timing of spawning. Thus it can be seen how the effective population size could, in certain circumstances, be very much smaller than the total number of individuals in the population.

If the effective population size (N_e) in the expression above is a very large number, then the allele frequency variance between successive generations will be very small. Conversely, allele frequency variance will be large when N_e is small. This means that natural fluctuations in allele frequency between generations will be much greater in small populations. We shall encounter this concept again when dealing with various aspects of genetics in small populations.

Allele frequencies in a population can therefore change over time owing to random genetic drift. However, they may also change owing to pressure of selection (one genotype or one allele survives better than others at a locus) or to patterns of migration or dispersal that may fluctuate in direction or in strength over time.

Population Structure Arise

The differentiation of a species into genetically different populations is a fundamental part of the process of evolution. For a number of physical or biological reasons, the distribution of a species may become fragmented. For example, during the last ice ages, aquatic species in the temperate regions of the northern hemisphere were driven south and many became restricted to local areas in southern habitats.

Where species were fragmented into different areas, many of these populations remained isolated for long periods of time in such refugia. Following the retreat of the last glaciation some 10 000–13 000 years ago, aquatic and terrestrial species went through the process of gradual recolonisation of habitat and this has provided plenty of opportunity for further fragmentation.

Once fragmentation has occurred, allele frequencies at most loci will be subject to random genetic drift in the fragmented populations. As we have seen, random genetic drift produces more rapid allele frequency change in small, compared with large, populations. If only a few individuals are involved initially in founding a population, then allele frequencies in the new population may be very different from the source population, and if numbers remain small, further allele frequency changes may be rapid.

In addition to random genetic drift, localised adaptation will occur involving selection at some loci for particular characteristics and these will cause further differences between populations. These loci under selection, and to a lesser extent those linked to them, can show higher patterns of differentiation between populations than neutral markers and are sometimes called 'outliers'.

These processes of random genetic drift and adaptation will tend to increase differences between populations, while migration or larval dispersal between populations will tend to reduce it. So the population differentiation we can identify today in a species is essentially the result of the historical interplay between the environment and the forces of dispersal, adaptation and random genetic drift.

Cytogenetic and FISH analysis

The differential diagnosis between 'true' SAA and hypoplastic MDS is often difficult because of a reduced bone marrow cellularity.1,2 Fundamental discriminant clues can be provided by trephine biopsy and by the identification of a chromosome aberration,3 that, however, has been reported in SAA patients too.

We report on five patients, aged 20 59 years, observed at our Institution between March 1989 and August 1995. Their main clinico-hematological characteristics. Cytogenetics and FISH results are report. At diagnosis an initial cytogenetic analysis showed a normal chromosome pattern in all the patients. All of them evolved into AML showing clonal karyotype defects described according to ISCN4 (trisomies and monosomies were considered clonal when present in at least two and three metaphases, respectively). We decided to employ FISH retrospectively in order to establish if any minor subclone, carrying the chromosome abnormality detected by cytogenetics at the time of AML transformation, was already present at diagnosis. In order to reinforce the interphase FISH analysis of small subpopulations of cells we decided to apply the probe discovering the clonal chromosome defect along with an additional probe used as an internal control (dual-colour FISH). All probes were commercially obtained.

The digoxygenated 5q31 DNA probe was applied to metaphase cells only along with the 5 probe. Its cut-off value was obtained by screening 1000 mitoses from 10 normal controls. We determined the one-sided 95% confidence limit using a binomial distribution for the proportion of metaphase cells that were normal or that carried the del(5)(q13q31). We detected the probability of discovering 0, 1, 2, 3 cells with del(5)(q13q31) by using the upper boundary of the 95% confidence interval.5 As in normal subjects, 3 was the highest number of cells lacking the 5q31DNA probe signal, probably because of a hybridization failure, the normal range for the cells with this karyotypic defect was fixed at 0.8%.

Monosomy 7 or trisomy 8 or other numerical defects were considered clonal by metaphase and interphase FISH when the percentages of the cells with such abnormalities were above the cut-off value determined for each numerical change. The cut-off value was obtained by screening 25 mitoses and 200 interphase cells from 10 normal controls. Monosomy 7 cut-off was determined by applying the 7 probe together with a 12 probe. We calculated the mean percentage of mitoses or cells showing two spots due to the internal control and only one spot due to the 7 probe plus three times the standard deviation. Monosomy 7 cut-off value for metaphase FISH was 8%; that for interphase FISH was 5%. Trisomy 4 and 8 cut-offs were obtained in the same way and were fixed at 0% for both the numerical defects when considering mitoses, at 2.3% and at 2.5%, respectively, when examining interphase cells.

Despite the high interphase FISH cut-offs for monosomies and trisomies making the identification of minor subclones marked by such abnormalities difficult, this technique succeeded in demonstrating that the clonal defect was already present at disease onset in a few interphase cells obtained from four of our patients. This result was confirmed by metaphase FISH, that detected similar percentages of clonogenic cells in the same patients, suggesting that, initially, the chromosome defect was really present in a small subpopulation of cells. Moreover, a careful review of the karyotypes, obtained at disease onset and considered normal on an initial conventional cytogenetics investigation, demonstrated that few mitoses of the same four patients carried the clonal abnormalities observed on leukemic transformation and retrospectively identified by FISH. These cytogenetics and FISH results might be explained by the very small size of the abnormal clone and not by the low mitotic rate with poor quality metaphases often observed in patients with hypoplastic marrows. In both situations, however, FISH plays an important role in identifying clonogenic cells that may easily escape cytogenetic detection.

An intriguing point in patients with hypocellular marrow is the differential diagnosis between 'true' SAA and hypoplastic MDS.1,2,3 In our patients FISH alone was unable to discriminate between true SAA and hypoplastic MDS, as the presence of a chromosome abnormality per se does not help to make a discriminatory diagnosis. However, karyotypic defects such as monosomy 7, trisomy 8 and sporadically trisomy 6 have been observed in SAA evolving into MDS or into AML.3 Therefore, an overlap between SAA and MDS may actually exist.2 This possibility is further illustrated by our patient No. 5, who showed an abnormality identifying a particular subgroup of MDS with a very low probability of evolving into AML, ie a del(5)(q13q31), accompanied by an increased marrow reticulin on trephine biopsy, a sign predictive of AML evolution.1 This case initially classified as hypo-MDS ultimately developed into AML-M4 88 months after initial diagnosis.

In conclusion, the present FISH study was made possible by conventional cytogenetics, the only reason that the abnormalities present in each case were known. Cytogenetics and FISH were complementary; the latter succeeded in discovering minor subclones at diagnosis and, demonstrating their expansion, it was able to predict AML transformation.

A Population Genetics Model

Now that you have seen what happens in a situation of no evolution (Hardy-Weinberg Equilibrium), let's develop a model of natural selection.

Remember that natural selection occurs if genotypes don't have equal fitness. We will model a situation in which natural selection is the ONLY form of evolution occurring (there will be no mutation, genetic drift, non-random mating, or gene flow--remember, these are the other forms of evolution, as introduced in the last lecture.) So for this model, we will assume (as we did for Hardy-Weinberg Equilibrium) infinite population size, no movement of individuals among populations, no mutation, and random mating. However, we will see what happens when the different genotypes for the trait differ in fitness. First, consider the fitnesses of genotypes. Remember that fitness is a measure of how well individuals with a certain genetic trait are expected to survive and reproduce. We will still consider a single gene with two alleles (A and a) and three genotypes (AA, Aa, and aa.) Natural selection will occur if individuals with different genotypes differ in fitness. So to understand and model natural selection, a first important point is to determine how to measure fitness.

Remember that natural selection occurs because individuals with some traits (those with higher fitness) survive and therefore reproduce more than do other individuals. The important feature here is reproduction--it is those traits that are reproduced most that necessarily become more common from generation to generation. So we will measure fitness based on the degree to which traits (in this case, different genotypes) reproduce. A first measure of the fitness of a genotype is absolute fitness: the number of offspring an individual with a particular genotype has on average. Absolute fitness could be studied in natural populations; one could determine that, for example, individuals with a certain genotype have on average six offspring.

A problem with this measure is that there is no way to determine, just looking at this number, whether having an average of six offspring means that this form has high or low fitness. If AA individuals have six offspring, Aa have three offspring, and aa have one offspring, then six is the best, and we would say it represents high fitness. On the other hand, if AA individuals have six offspring, Aa have twelve, and aa have thirty, then we would say six represents very low fitness. The important point to note here is that whether a particular absolute fitness is high or low depends on the fitnesses of other genotypes in the population. So we will develop a second measure of fitness that accounts for this.

A second measure of the fitness of a genotype is relative fitness: the degree to which individuals with a particular genotype reproduce in proportion to individuals with other genotypes. Typically, relative fitness is measured with respect to the genotype that has the highest fitness. The genotype with highest fitness has a relative fitness value of one; the other fitnesses have relative fitness values that are fractions of one, telling us how well these individuals reproduce with respect to the individuals of the genotype that reproduces best.

At this point, you should have all the information you need to work all the problems in Chapter VI of your lab manual, and you should try to start working on them. Here are some tips on how to work the problems: Your first step should be to determine whether you're working a problem for a population in Hardy-Weinberg Equilibrium or a problem in which natural selection is occurring. Remember that the difference between the two models is in whether or not different genotypes have different fitness. If the genotypes all have the same fitness (all survive and reproduce equally well) that should tell you you're looking at a Hardy-Weinberg Equilibrium problem. If the genotypes differ in fitness (if they don't survive and reproduce equally well) that should tell you you're looking at a natural selection problem.

If you have a Hardy-Weinberg Equilibrium problem there are two places the problem could start. You could be given allele frequencies to start out or you could be given genotype frequencies to start out. So your next step in solving the problem is simply to figure out, and write down, which you've got: a llele frequencies or genotype frequencies.

In a Hardy-Weinberg Equilibrium problem remember that if you have an allele frequency, you can figure out the other allele frequency (since p+q=1) and you can figure out any genotype frequency (Freq(AA)=p2, Freq(Aa)=2pq, Freq(aa)=q2.) So if you're given an allele frequency, use it, and get whatever information the question asks for. If you're given a genotype frequency rather than an allele frequency, your first step should be to calculate an allele frequency, because once you have the allele frequencies you can figure out anything else the question might require. To calculate an allele frequency from a genotype frequency, you need the genotype frequency of a homozygote (that is, either AA or aa.) Once you have that, take the square root to get the allele frequency of the allele that makes up that genotype.

If you have a natural selection problem then there are three places the problem could start, and you need to figure out which one you have. As with Hardy-Weinberg problems, things go more smoothly if you have allele frequencies, so your first step is to get an allele frequency (if you're not already given one.) Here are the three places a natural selection problem could start, and what your first step should be for each possible starting place (once you've completed the first step, you should be able to use the natural selection formulas to work the rest of the problem--it is often the first step that is hardest.)

1. You could be given an allele frequency in gametes at the start of a generation. If you are, use the allele frequency to determine the

other allele frequency (remember p+q=1) and use the fitness information you are given (be sure you take this and calculate relative fitnesses if you're not already given them) and use the formulas you have to get the information required by the question.

2. You could be given a genotype frequency in zygotes. If you are, remember that genotype frequencies in zygotes are p2, 2pq, and q2, just like in Hardy-Weinberg Equilibrium. If you have the genotype frequency of a homozygote in zygotes, take the square root to get the allele frequency that was present at the start of the generation, and proceed as you did in the first case where you were given the allele frequency directly.

You could be given a genotype frequency in adults. If you are, then you are in a situation where the population is NOT in the proportions of Hardy-Weinberg Equilibrium. To calculate the allele frequencies in these adults, which are also the allele frequencies in the gametes that will start the next generation, use the formula for the allele frequencies: Freq(A)=Freq(AA) + (1/2)Freq(Aa.) Once you have the allele frequencies that start the next generation, proceed as you would in the first case where you were given the allele frequencies directly.

When e-mailing me problems, show your work as clearly as you can. You can use the following ways of representing things to make typing the formulas (which is tedious, as I know, just having typed out a bunch of them for this lecture) easier: To represent p squared (p2) type p^2. To represent w bar, the average population fitness, type wbar.

To represent things with subscripts, like the absolute and relative fitnesses, you don't have to put the subscripts as subscripts--it's OK to type:

wAA for wAA

It will probably help to work stuff with pen and paper first, then type it out. Get into the habit of writing down all formulas and steps clearly and legibly--you'll make many fewer mistakes, you'll be able to go back and study from what you've done, and I'll expect you to write down all formulas and steps clearly and legibly on exams so you'll do better if you get into the habit of doing so now. Finally, from time to time look at the questions and the answers and think about what you'd really expect to happen in the situation you're given, through natural selection. Do the numbers you're getting makes sense? Can you see how they might relate to a real situation? The goal, really, is to be able to translate evolution into math and then the math back into evolution--don't get so bogged down in algebra that you forget what you're really trying to study.

Now that you've learned the basic model, let's consider some more aspects of natural selection as it affects situations such as those you have modeled: natural selection occurring because of different fitnesses of genotypes for a single gene with two alternate alleles in a diploid, sexually reproducing population.

Depending on the traits being studied, and the environment, fitness relationships among genotypes vary. They also depend on just how the traits are coded--whether one allele is completely dominant to the other (recessive) allele or whether there is some form of codominance or incomplete dominance such that the heterozygote has a different phenotype from either homozygote. (If these terms--dominance, recessiveness, incomplete dominance, codominance--are not completely familiar to you, be sure to review some genetics before proceeding!)

An important point to remember is that fitness depends on the phenotype--the appearance, structure, function of an organism. If two different genotypes result in the same phenotype (as when one allele is dominant) then both of those genotypes will have the same fitness--they look, act, function just like each other, so they survive and reproduce just like each other. Here are the possible ways the fitnesses of different genotypes could be related to each other: If one allele is completely dominant to the other, recessive, allele, so that the heterozygote has same phenotype as the dominant homozygote, there are two ways fitnesses could be related:

1. The recessive phenotype could have highest fitness (it would have a relative fitness value of 1; the dominant homozygote and heterozygote would have fitness values that are equal to each other and that are less than 1.)
2. The dominant phenotype could have highest fitness (in this case, the dominant homozygote and the heterozygote would both have relative fitness values equal to one, and the recessive homozygote would have a relative fitness value equal to something less than 1.)

If neither allele is completely dominant (in a case of codominance or incomplete dominance), so that the heterozygote has different phenotype from either homozygote, there are three ways that fitnesses could be related to each other:

1. Fitness codominance occurs when the heterozygote has fitness that is intermediate between the fitnesses of the two homozygotes (one homozygote has highest fitness, the heterozygote has lower fitness than it, and the other homozygote has still lower fitness.)
2. Heterosis occurs when the heterozygote has higher fitness than either homozygote (the heterozygote would have a relative fitness value of one, the two homozygotes would have fitness values of less than one.)

3. Underdominance (negative heterosis) occurs when the heterozygote has lower fitness than does either homozygote (one homozygote would have a relative fitness value of one, the other homozygote might also have a fitness value of one or, more likely, would have a somewhat lower fitness value, and the fitness value of the heterozygote would be even lower.)

You can use the formulas you've learned to calculate the effect of natural selection over one or two generations for any of these situations. If you repeated calculation of these formulas for 50 or 100 generations (I don't recommend doing this by hand!) you could learn what happens in each of the situations, in the long run. It turns out that the different situations listed above do differ in the expected result of natural selection. To learn about these, you will complete a computer exercise to study the different situations. In this assignment, you will look at specific instances of the forms of natural selection and of one other form of evolution, genetic drift. These specific instances illustrate general patterns--for example, what you learn for the specific case of heterosis you study will show you patterns that are generally true for heterosis. Your goal will be to learn these general patterns.

laboratory techniques

Novel laboratory techniques are often the driving force behind new ideas in science. Certainly the use of cytogenetic studies for clinical testing literally exploded in the 1960s and 1970s because of three new techniques:

(1) Peter Nowell discovered that the plant lectin phytohaemagglutinin could stimulate lymphocytes to grow in culture;

(2) TC Hsu used hypotonic shock to prepare chromosomes so that they could be counted accurately;

(3) T. Caspersson and Marina Seabright developed chromosome banding techniques that allow accurate pairing and identification of structural abnormalities.

In the 1980s the techniques of molecular biology were applied to cytogenetic preparations.

We call this "hybrid" technology molecular cytogenetics, and it is transforming the way we study chromosomal changes in humans. It has improved the detection of, indeed in some cases defined, microdeletion syndromes. De novo derivative chromosomes in patients with chromosome imbalances can now be accurately characterized: the clinician now knows the identity of the partial monosomy and trisomy in these patients, and the molecular geneticist has accurate information from which to find genes relevant to the phenotype. The chromosomal origin of supernumerary markers can be identified and we now have a real potential of predicting phenotype for marker carriers, benign or adverse. With molecular cytogenetic techniques, we can now detect some chromosomal abnormalities in nondividing cells with interphase nuclei. Standard cytogenetics requires actively dividing cells with metaphase nuclei. Some constitutional chromosomal abnormalities, and phenomena such as mosaicism and chimerism can be addressed without growing cells in culture or in cell types that do not adapt well to tissue culture.

Potentially, chromosomal aneuploidy studies (*i.e.* studies looking for abnormalities in chromosome number) can be done more quickly. Preserved, rather than fresh, tissue can be investigated for retrospective studies. We can ask questions of chromosome organization, its relationship to gene expression and tissue specificity in ways that were not possible a few years ago.

Does it replace standard cytogenetics? No, molecular cytogenetic techniques focus on specific chromosomes, chromosome regions, and unique DNA sequences or genes. Standard cytogenetic studies, including high resolution analysis, allow us to survey the whole genome for abnormalities of chromosome number or structure. But as complementary studies or for certain targeted abnormalities, molecular cytogenetic techniques expand our capabilities for making more accurate and refined cytogenetic diagnoses, both for constitutional abnormalities and acquired chromosomal changes in cancer cells.

Basic techniques

The most commonly used molecular cytogenetic technique is fluorescence in situ hybridization, or FISH. FISH is usually applied to standard cytogenetic preparations on microscope slides, but it can be used on slides of formalin-fixed tissue, blood or bone marrow smears, and directly fixed cells or other nuclear isolates. The basic principle of the method is that single-stranded DNA will bind or anneal to its complementary DNA sequence. Thus, a DNA probe for a specific chromosomal region will recognize and hybridize to its complementary sequence on a metaphase chromosome or within an interphase nucleus. Both have to be in single-strand conformation, therefore the DNA probe and the target DNA must be denatured, usually by heating them in a formamide-containing solution.

The probe is hybridized to the target DNA under conditions that allow the DNA to reanneal in double-strand form. Added to the hybridization mixture is an excess of repetitive sequence DNA to block non-specific binding of the probe to the target. After hybridization is complete (often after 2-18 hr at 37 C), the slides are washed in formamide-saline citrate solutions to remove excess or non-specifically bound probe. To detect the location of the probe on the target DNA, the probe DNA can be directly labeled with a fluorescent tag. It can also be chemically modified by the addition of hapten molecules (biotin or digoxigenin) that can then be indirectly fluorescently labeled with immunocytochemical techniques. The target DNA is counterstained with another fluorochrome of a complementary colour. The probe DNA can be observed on its target by using a fluorescent microscope with filters specific for the fluorochrome label and the counterstain. Special filters have been developed to allow simultaneous visualization of several fluorochromes.

Digital cameras designed to detect low light level emissions and computer imaging are used to increase the sensitivity of probe detection. Because fluorescent dyes are subject to photobleaching (fading), the preparations are not permanent and must be stored away from light. Use of an antifade solution (phenylenediamine) has improved the capacity to observe and document fluorescently labeled samples.

Three different types of probes are typically used in clinical FISH studies:

1. Probes that bind to specific chromosome structures typically recognize repetitive DNA sequences, such as within centromeres (alpha satellite DNA) or telomeric sequences. Within the repeat sequence are nucleotide patterns that are unique for specific chromosomes.

   There are now available centromeric probes that can identify most of the individual chromosome homologues. Similarly, there are probes that specifically recognize the telomeres (chromosome ends) of the long and short arms of many of the homologues, and more are rapidly being developed. Most alpha satellite centromeric probes give a large, bright signal and are useful for both chromosome identification in metaphase preparations and chromosome enumeration in interphase nuclei. Chromosome-specific telomere probes can be used for the above, for detection of cryptic translocations, and to define interstitial and terminal deletions.

2. Unique sequence probes hybridize to single copy DNA sequences in a specific chromosomal region or gene. In clinical cytogenetics these probes are usually referred to as cosmids, named for their cloning vector. These are the probes used to identify the chromosomal critical region or gene associated with microdeletion syndromes. On metaphase chromosomes, they hybridize to each chromatid, usually giving two small, discrete signals per chromosome.

3. Whole chromosome paints are cocktails of unique sequence probes that recognize the unique sequences spanning the length of a particular chromosome. At metaphase, both chromosome homologues are "painted" or fluoresce brightly. Among other applications, paint probes are used to define the chromosomal origin of derivative segments on translocation chromosomes or supernumerary markers.

A new nomenclature to describe FISH findings has been introduced by the standing committee for the International System on Cytogenetic Nomenclature.

It is designed to be compatible with standard cytogenetic nomenclature and to provide precise information about FISH results. Many laboratories will be incorporating the new nomenclature into their reports, together with text to explain their findings.

The Genetic Drift Newsletter is not copyrighted. Readers are free to duplicate all or parts of its contents. The Genetic Drift Newsletter is published semiannually by the Mountain States Genetics Network for associates & those interested in Human Genetics.

In accordance with accepted publication standards, we request acknowledgement in print of any article reproduced in another publication. The views expressed in the newsletter do not necessarily reflect local, state, or federal policy.

Fish Applications in Cancer Cytogenetics

FISH analysis can identify selected acquired genetic changes in nondividing (interphase) cells and can detect masked translocations on metaphase chromosomes. However, FISH is informative only for the specific probes used, while cytogenetic analysis can detect a wide range of abnormalities and provides both diagnostic and prognostic information for patients with a variety of malignancies. The two techniques are complementary, and FISH has proven a valuable adjunct to cytogenetic analysis in the following ways:

Fish Analysis of Interphase Cells

This application requires that a probe be available for the abnormality in question and that the laboratory has established normal background ranges for that specific probe. Interphase cells may show lack of signal or may demonstrate multiple signals because of various technical artifacts. It is important to establish the range of signals observed in normal cells for each probe used, in order to determine what this background rate is.

Given that these conditions have been met, FISH analysis is very useful for:

Identification of Abnormalities in Specimens with a Low Mitotic Index

Diseases such as the chronic lymphoproliferative disorders are characterized by a very low proportion of mitotically active abnormal cells. In these cases traditional cytogenetic analysis is more likely to identify the karyotype of the normal dividing cells, and may miss the rare abnormal cell in mitosis. FISH on interphase cells from these patients is often helpful in detecting abnormalities when there is a high suspicion of a specific abnormality associated with the disease in question-for example, trisomy 12 or a deletion of 13q14 in patients with CLL. In cases of cytogenetic culture failures, FISH may also be helpful if an abnormality, such as the t in CML, is suspected.

Single Cytogenetically Abnormal Cell is Detected

Because cytogenetically abnormal cells may arise as artifacts of the tissue culture or slide making processes, the determination of clonality (*i.e.* that all of the abnormal cells originated from a single aberrant cell) requires that multiple cells share the same aberration. When only a single abnormal cell is observed in the course of a cytogenetic analysis, FISH on interphase cells may be used to confirm the presence of an abnormal clone, if the appropriate probe is available and if the laboratory has established normal ranges for that probe.

Treatment and Detection of Minimal Residual Disease (MRD)

If a cytogenetic abnormality has been identified at diagnosis and a FISH probe is available to detect that abnormality, FISH may be used to screen interphase cells in subsequent studies. However FISH alone is not sufficient for monitoring, because cytogenetic analysis may also detect additional acquired changes, which may have important prognostic significance, or may uncover the development of secondary disease with a different set of abnormalities.

The sensitivity of cytogenetic analysis of mitotic cells and FISH analysis of interphase cells is roughly equivalent-*i.e.* a population of about 5% abnormal cells can be detected by either approach-although FISH allows a greater sampling of cell types since mitotic cells are not required.

The ability of FISH to be used to screen hundreds of interphase cells may increase sensitivity for the detection of MRD, especially when samples can be enriched for the presumptive abnormal population. However, the clinical significance of abnormal cells which are not actively dividing is not established and may vary with both the disease and the therapy used to treat it.

Fish Analysis of Metaphase Chromosomes

This technique employs the use of whole chromosome paints, which are now commercially available for every chromosome, or cosmid probes on metaphase spreads prepared in the course of cytogenetic analysis. This applications allows:

Variant Translocations or in Cytogenetically Normal Cells

The application of molecular techniques to cancer diagnosis has taught us that some patients may have apparently normal chromosomes in their neoplastic cells but nevertheless have acquired genetic changes which are detectable only by molecular methods. FISH analysis with cosmid probes is useful for detecting masked bcr/abl rearrangements in CML, or pml/rara rearrangements in suspected APL. These cosmid probes may also be used for interphase cell analysis, with sensitivity similar to southern blotting, but with results more rapidly obtained.

Clarifying Complex Rearrangements and Breakpoints

Whole chromosome paints or appropriately located cosmids may be hybridized to metaphase chromosomes to identify the origins of complex marker chromosomes, or to identify cryptic translocations which may appear as simple deletions.

For example, some cases of myeloid leukemias with apparent deletions of 11q23 may actually have cryptic translocations, often involving chromosome 6. Delineation of disease groups and prognosis may hinge on this information.

The selective application of FISH techniques to cancer cytogenetic analysis has the ability to increase the resolution, sensitivity and informativeness of each analysis. In order to be more widely applied, there must be a greater selection of commercially available probes, and issues of quality control and quality assurance still need to be addressed.

Fish for Gene Amplification

Role of Gene Amplification in Cancer

Gene amplification is a characteristic of cancer cells that allows increased production of specific proteins used for acquisition and maintenance of the malignant phenotype. Amplification of certain oncogenes has an important role in the progression of many tumors. Detection of such amplifications may be of diagnostic and prognostic importance.

The HER-2/neu or erbB-2 oncogene codes for a 185 Kd transmembrane oncoprotein (p185). Approximately 25-30% of breast and ovarian carcinomas have amplification of this gene. Studies have demonstrated that HER-2/neu is amplified and/or overexpressed in 10-30% of invasive breast cancers and in 40-60% of intraductal breast carcinomas.

Amplification and overexpression of this gene were found to be associated with rapid proliferation, low estrogen receptor content, and high grade of ductal carcinomas, which suggests that this oncogene plays an important role in the progression of breast cancer.

Amplification or overexpression of the HER-2/neu gene was shown to be an indicator of poor prognosis in node-positive breast cancer. In node negative breast cancer patients, the subgroup for which accurate prognostic factors could make a significant contribution to treatment decisions, the prognostic value of HER-2/neu amplification or overexpression has been controversial.

The technical difficulties in using standard immunohstochemical staining (IHC) and Southern blot analysis may explain not only the differences in the observed clinical-genetic correlation, but also the varied frequency of HER-2/neu overexpression reported in tumors.

Fluorescence in situ hybridization (FISH) has been used to demonstrate gene amplification in human breast carcinoma cell lines in both interphase and metaphase cells. FISH appears to be a useful technique as it can overcome many of the inherent limitations of other techniques and can provide the most reliable and accurate data. HER-2/neu amplification appears to be a strong indicator of breast cancer recurrence in women under 40. This gradient of risk demonstrated by FISH with a DNA probe and the more consistent recent findings of the prognostic importance of amplified HER-2/neu is evidence of probable superiority of this technique compared to standard Southern blot or IHC.

5

Genetic Aquaculture Resources

Genetic resources are the foundation on which species, stocks and genetically-improved strains are based. At the species level, more aquatic animals are being farmed now than ever before.

Although the common carp, Cyprinus carpio and goldfish, Carasius auratus, were domesticated several thousand years ago into a variety of shapes and colours, most of the farmed fish today are very similar to their wild relatives. Improvements in our knowledge of artificial reproduction, reproductive biology, early larval rearing (training series) and basic genetics have recently allowed fish breeders to improve genetically species such as rainbow trout, coho and Atlantic salmon, channel catfish, Nile tilapia, as well as common carp.

The culture of several important species still relies on the collection of brood stock or seed from natural populations. Perhaps the most important group of species whose culture is dependent on natural populations is the marine shrimp Penaeus spp. Shrimp farming in South and Central America stock production ponds with wild-caught larvae. Hatcheries that produce shrimp larvae exist in Asia and the Americas, but the broodstock are generally collected from the wild. Culturists recognise the problems associated with this harvest of wild resources and are taking steps to domesticate marine shrimp. Other culture systems dependent on wild resources include milkfish in the Philippines, yellow tail in Japan, and eel in Asia and Europe.

However, simply having a domesticated species or genetically improved species is not sufficient to guarantee optimum production from an aquaculture facility. In addition to proper husbandry, *i.e.*, water quality, nutrition, health etc., broodstock must be managed to ensure conservation of genetic resources, to maintain the desirable characters of the farmed species and to avoid problems of inbreeding.

Aquatic Resources and their Utilization

More than two-thirds of the world's surface is covered by water and the total yearly production of organic material in the aquatic environment has been estimated at about 40 000 million t . Tiny microscopic plants, the phytoplankton, are the primary producers of organic material using the energy supplied by the sun.

This enormous primary production is the first link in the food chain and forms the basis for all life in the sea. How much harvestable fish results from this primary production has been the subject of much speculation.

However, there are great difficulties in estimating the ecological efficiency, *i.e.*, the ratio of total production at each successive trophic level. Gulland (1971) reports a range from 10 to 25 % but suggests 25 % as the absolute upper limit of ecological efficiency; for example, not all of the production at one trophic level is consumed by the next.

Ecological efficiency also varies between levels, being higher at the lower levels of the food chain with smaller organisms using proportionally more of their food intake for growth rather than for maintenance. Diseases, mortality, pollution, etc. may also influence ecological efficiency. Since production is greater in the early stages of the food chain, the potential catch is also greater if harvesting is carried out at these stages.

Up to 1970, the world catch of marine fish continued to rise at an overall rate of 6 percent per year, according to FAO statistics. Great optimism was expressed by various authors who estimated the potential world catch to be somewhere between 200 million t/year to 2 thousand t/ year ; most of this wide variation being due to uncertainties concerning the trophic level at which the harvest would be taken.

The yearly increase in catches has slowed down since 1970, and the total catch reached a peak of 100 million t in 1989. Since then it has started to drop as a number of fish stocks have begun to collapse, in many cases due to overfishing. However, a slight upward trend is noticed for 1992 and for 1993 world catch is estimated to reach 101 million t. While total catch has started to decline since the peak in 1989, the catch from developing countries as a group is still increasing and since 1985 has exceeded that from developed countries.

Thus in 1992 little more than 60 % of the total world catch was taken by developing countries, and it is estimated that this figure will increase to 66% in 1993. This also means that an increasing part of the world fish catch is taken from warm tropical waters.Are we then reaching the limits of production from "wild" aquatic resources now or do the optimistic predictions from the 1970s still hold? The answer to this question is not only in the affirmative, but for many resources the limit was reached decades earlier than the peak in global landings .

A combination of factors has helped to mark the depletion of many conventional resources. One of these is that continued investments in fishing fleets throughout the world has meant that although catch rates and abundance of high value fish species have often declined, the overall level of fishing effort has increased so that roughly similar levels of landings are being taken at much greater cost to many fishing nations.

The real problems with decreasing fish stocks are familiar. First there is "the tragedy of the commons" - whatever lacks a known owner, whether buffalo or fish - which everyone will race to exploit and ultimately destroy.

The next problem which can be identified is the exceptionally poor management of the aquatic resources. What has been done has been too late and too little. The 1982 Law of the Sea, which extended the territorial seas from 12 to 200 miles, gave the coastal States an opportunity to take a protective interest in their fishing grounds.

Instead, many of them rushed to plunder the resources by offering generous subsidies and tax relief for new vessels. Also, the much used quota-system is subject to severe criticism. Often, the net result is increased fishing and increased waste, as perfectly good fish are thrown overboard if quotas are already reached.

Many fish stocks (such as pollack, haddock and halibut off New England) are now considered "commercially extinct"; that is, there are now too few fish to warrant catching. From an initial stage of under-utilization the fishing passes through a phase of rapid expansion until the limit of the resource is reached. This is then followed by a period o overfishing with high fishing effort, but reduced catches until finally - and hopefully - a phase of proper management is reached.

Details on resource management are beyond the scope of this book, but should include the concept of sustainability, environmental aspects and responsible fishing.

However, in an FAO publication it is stated that change from a focus on short-term development of fishing fleets to proper management is a necessary, but insufficient condition for sustainable development. In the same report it is further stated that "Sustainable Development" as promoted at the United Nations Conference on Environment and Development (UNCED) in 1992 cannot be achieved under open-access regimes, whether these are within or outside national territorial waters.

In contrast, the world aquaculture production inclusive of aquatic plants has steadily increased over the last decade totalling 19.3 million ton in 1992, almost half of this (49% is produced in marine aquaculture, 44% in inland aquaculture, and the rest in brackish environment. About 49% of world aquaculture production are fish.

Production of aquati plants is increasing rapidly and reached 5.4 million t in 1992, while smaller increases if production of molluscs and crustaceans are seen.

The total value of the aquaculture production is estimated to more than $US 32.5 billion in 1992. To summarize, it can be said that further increases in supply of fish can be expected from better utilization/ reduction of losses and further expansion of aquaculture.

Table shows the breakdown of world fish production.

Table : Breakdown of World Ffish Production (percentage of world total in live weight)

Year	*For human consumption*			*Other purposes*		
	Total	*Fresh*	*Freezing*	*Curing*	*Canning*	*Animal Feed*
1982	71.1	19.4	25.3	12.8	13.6	28.9
1992	72.8	27.0	24.1	9.3	12.4	27.2

Table shows relatively modest differences in the breakdown of the fish production during the decade 1982-92. However, there was a significant increase in fresh fish consumption. Total fish for human consumption increased by 1.2% while fish used for curing and canning continued to decrease. In value terms, fishery exports reached an estimated $US 40.1 billion in 1993.

Exports of fish and fishery products from developing countries continued to increase reaching a total value of $US 19.4 billion in 1993. In the same year exports from developed countries dropped by 5% to an estimated total value of $US 20.7 billion. Developing countries recorded an increasingly positive trade balance in fish trade, which reached $US 12.7 billion in 1993 .

It should be noted that Table does not give a true picture of the amount of fish available for human food. An enormous amount of fish is wasted due to discards on board or post-harvest losses during processing and distribution. It has been estimated that the global amount of discards is in the range of 17-39 million t/year with an average of 27 million t/year . It has been further estimated that the total post-harvest losses in fish products are about 10 % (James, D., personal communication 1994). These high losses are mainly due to problems of fisheries management, and lack of proper technology and of economic incentives.

Cell Culture in Aquaculture

Development and characterization of cell lines from selected finfish and shell fish was undertaken at CIFE, Mumbai and a cell culture system was developed from the fish Lates calcarifer (Sea bass). At CUSAT, the cell lines developed from eyestalk sub-cultured and made available to develop such cell lines as and when required.

The cells shown differentiation in to a proliferating bunch of cells and subsequently got detached as a lump of cell aggregates. The development of cell lines from seabass, Lates calcarifer was attempted at CAH Abdul Hakeem College, Melvisharam and CUSAT. The kidney cells shown fibroblastic and formed a monolayer. The cells were were deposited at NCCS.

Growth Promoting Factor(s)/Signal Transduction

A collaborative project was undertaken on identification, isolation and characterization of growth promoting and differentiation factor(s) from perivitelline fluid of developing embryos of the Indian horseshoe crab at Agharkar Research Institute, Pune, National Centre for Cell Science (NCCS), Pune and National Institute of Oceanography (NIO), Goa. Active fraction identified shown significant enlargement and enhanced compartmentalization of the heart in cardiac development and promoting activity. In another project on large-scale cultivation of amoebocytes in vitro from Indian horseshoe crab implemented by NCCS and NIO, regeneration of gill lamellae of the Indian horseshoe crab was successful in amoebocyte production.

A molecular signal intervention mechanism was developed as an alternative to eyestalk ablation at the Departments of Zoology and Biochemistry, University of Madras. A specific inhibitor that intervenes in this pathway has been shown to reduce the molt cycle of the prawn and reduces the turnover time and rearing cost and maximizes the economic returns. By decreasing the molt cycle duration, it can accelerate the growth rate of the prawn resulting in early maturation.

Studies on feed supplementation of highly unsaturated fatty acids (HUFAs) and immunostimulants for disease resistance in freshwater prawn was carried out at the Department of Aquaculture, College of Fisheries, Mangalore.

HUFAs namely Eicosapentaenoic acid (C20: 5n-3) and Docosahexaenoic acid (C22: 6n-3) are important for larval growth. The supplementation of PUFA incorporated feed could enhance growth, survival and disease resistance in freshwater prawn. A project on prototyping of raceway based third generation shrimp production technology was implemented at the Fisheries Biotechnology Centre, Fisheries College and Research Institute, Tuticorin.

The protocol has been developed to increase survival of post-larvae. This in-door shrimp farming facility will be eco-friendly, economically viable and technically useful for large-scale adoption in aquafarming systems in India.

The work on development of Artemia franciscana culture in the solar salt ponds and processing of Artemia cyst and biomass was undertaken at Centre for Marine Science and Technology, Manonmaniam Sundaranar University, Rajakamangalam, Tamil Nadu. Artemia production was initiated in condenser ponds and inoculated in culture ponds and the efforts are being made to utilize the abandoned area of the saltworks for Artemia cultivation. Studies were undertaken at NIO, Goa to isolate, purify and characterize new natural fluorescent dyes and fluorescent molecules from tissues of sea cucumber Holothuria scabra and to test them as bioactivities compounds for therapeutic and industrial applications.

At Kerala Agriculture University, Kochi, attempts were made to isolate bioactive substances such as antibacterial, antiviral and anticancer agents from the accessory nidamental gland and ink of cuttlefish (Sepia pharaonis ehrenberg). An antitumour peptidoglycan was isolated from cuttlefish ink and the crude preparation showed antitumour activity against mice following intraperitonial administration. The peptidoglycan was made up of protein and uronic acid rich with polysaccharide fraction and was found to be made up of five amino acids. The screening of marine Acinetobacter genospecies for production of biosurfactant/s, purification and antimicrobial activity was undertaken at Pune University. Untapped natural sources were screened for novel biosurfactant/s and bioemulsifier-producing microorganisms using Acinetobacter strains, which showed emulsification of toluene, diesel, hexadecane and crude oil, haemolysis of human RBC's and anionic biosurfactant/s production.

At the department of biotechnology - CUSAT, the work on molecular cloning was undertaken to develop a technology for production of alkaline protease enzyme using marine fungus, Engyodontium alba. The optimal conditions were studied for enzyme production using solid-state fermentation of wheat bran as the solid substrate. The enzyme was found relatively stable for one year. At NIO, Goa, fungi from deep-sea sediments were explored as new sources for protease. Fungal growth and protease production at low temperature and elevated hydrostatic pressure was studied to understand their adaptations to extremophilic conditions.

Seaweed Cultivation

At CSMCRI, Bhavnagar, the project was undertaken on scale-up cultivation and processing of phycocolloid seaweeds (Gelidiella acerosa and Eucheuma). Multilocational trials undertaken to validate different culture methods for large-scale cultivation of Eucheuma proved that the raft cultivation method consistently yields better harvest. This method has been successfully demonstrated in the Gulf of Mannar, Tamil Nadu coast as well as at Okha, Gujarat coast. High quality agar was extracted from Gelidiella acerosa and Gracilaria sp. The Gracilaria agar had very high gel strength (1200 g cm-2) with low gelling temperature. This agar will be useful in bacteriological and molecular biology applications.

The nitrifying bioreactor technology developed at CUSAT paves the way for transforming the open prawn/shrimp larval rearing systems to closed re-circulation systems. The reactor, on properly activating and deploying in the hatchery system, can take care of the entire water quality problems. Apart from nitrification this duel capability lowers down the biological oxygen demand, total bacterial population along with stabilization of Vibrio population. The training programme for fisheries scientists was undertaken at three institutes namely Central Institute of Freshwater Aquaculture (CIFA), Bhubaneshwar, Central Institute Fisheries Technology (CIFT), Cochin and Central Institute of Fisheries Education (CIFE), Mumbai to train in service fisheries scientists on aspects of modern biology, health management and conservation of resources etc. Trainees are being given both theoretical hands-on training and independent project work on molecular biology aspects relevant to fisheries topics. After successful completion of the training programme, the trainees have started working in the area of molecular biology in marine sciences at their institutes.

New Programme Support

The programme on aquaculture and marine biotechnology has been identified for support after a detailed discussion under the Task Force on Aquaculture and Marine Biotechnology. Two proposals from the UAS, College of Fisheries, Mangalore and the Cochin University of Science and Technology (CUSAT), Cochin have been supported for attention on brood stock and larval health management of black tiger shrimp and freshwater prawn and screening for potential anticancer agents from marine Cnidarians, bacteria and fungi. Genotypic characterization and sequencing of genes of interest of marine bacteria, fish and shellfish immunology and screening for useful natural products from microorganisms, capacity building through hands-on training programmes for scholars and technicians on molecular techniques for detection of pathogens and health management in shrimp hatchery system, microbial taxonomy, crustacean immunology and screening and characterization of marine microbes for useful products etc. are also the part of the programme.

Shrimp Genomics

Shrimp (Penaeus monodon) is a very important species for the country from the commercial viewpoint. Unfortunately, the species is not domesticated and the harvests are mainly from the wild. The reproductive biology is complex and the disease burden very high, especially viral diseases.

There are populations, which are naturally resistant to viral attack and could be tapped to our advantage. The network programme on shrimp genomics has been developed covering the aspects *viz.* functional genomics in disease resistance of Penaeus monodon, gene regulation of maturation and breeding in P. monodon, microsatellites and physical mapping of P. monodon and genomics of Penaeus indicus for comparative genomics.

Technologies/Products

(a) Development of a heat killed whole cell vibrio vaccine using a virulent strain of Vibrio harveyi was developed at College of Fisheries, Mangalore. This has shown immune response in shrimp (Penaeus monodon). The technology has been transferred to Mangalore Biotech Lab.

(b) A polyclonal antibody based immunodiagnostic kits for detection of different bacterial pathogens in finfish and shellfish has been developed by CIFE, Mumbai. The technology has been transferred to GlaxoSmithKline Pharmaceuticals, Mumbai without any modifications for commercialization. The kits are useful as on-farm, rapid diagnostic kits to detect bacterial and viral diseases of finfish and shellfish.

(c) To overcome the problem of viral diseases in shrimp aquaculture and for simultaneous detection of WSSV and MBV, highly sensitive method "Combi kit – WSSV, MBV", a test kit, for specific detection of pathogens is developed by the Department of Microbiology, UAS, College of Fisheries, Mangalore. The test kit can detect all genotypes of WSSV and can be used to detect the presence of virus in shrimp, pond water, sediment, carrier animals etc. The technology has been transferred to Mangalore Biotech Laboratory for packaging and commercialization.

(d) A polyclonal antibody-based immunodiagnostic assay for the detection of white spot syndrome virus - a simple diagnostic test kit developed in joint collaboration with the Centre for Biotechnology, Anna University and C. Abdul Hakeem College, Melvishram was transferred to an aquaculture-based company (Poseidon Biotech, Chennai)

This book has been planned and written in simple and easy language, and covers the almost courses of all Indian Universities.

Aquaculture Technology

The technology employed in aquaculture has developed over many centuries but has done so more rapidly during the last half century. Aquaculture systems, and the technology used, vary from very simple systems, used for family ponds in tropical countries, where production is for domestic consumption, to high technology systems, such as intensive closed systems, like those used for rearing stripped bass.

Herbivorous and filter feeding fish, reared mostly in simple systems of small freshwater ponds, however, account for about half of global aquaculture production.

Much of the technology used in aquaculture is relatively simple, amounting to small modifications that improve the growth and survival rates of the target species, such as providing additional food, adding seed animals collected elsewhere, managing water exchange to maintain adequate oxygen levels, and protecting the stock from predators. Greater understanding of complex interactions between nutrients, bacteria and cultured organisms, together with advances in hydrodynamics applied to pond and tank design, have enabled the development of closed systems. These have the advantage of isolating the aquaculture systems from natural aquatic systems, thus minimizing the risk of disease or genetic impacts on the external systems.

Developments in engineering, some learnt from offshore oilrig construction, increase the possibilities for offshore aquaculture using robust cages. Sea ranching, the release of young fish into the wild to improve the harvest in capture fisheries, has also made a start but its long term viability is still to be assessed. Major advances are also being made in the technology of the production of aquafeeds, which generally require the combining of a large number of ingredients into very small feed pellets.

Aquaculture Engineering

Aquaculture draws on well-established engineering fields for most of the design and construction needs of its production facilities. Building earth ponds is similar to building roads - a knowledge of the characteristics of soils and the limits of safe design are the basis of good construction. Similarly, the buildings used for hatcheries and other support activities are no different from those common in the housing, agricultural and commercial sectors. Sometimes ponds are lined with plastics or other impermeable materials, and here the techniques are similar to those for civil structures such as potable water reservoirs or sludge tanks.

The design and installation of water control gates, including in unstable soils, benefits from the long experience in this field in the agriculture and irrigation sectors. An understanding of hydrodynamics allows ponds and tanks to be built with good water circulation, oxygen mixing and without 'dead spots' where sediments might accumulate and cause health problems to fish.

For installations in the sea, the situation is somewhat different and many of the important engineering solutions, such as for fish cages or suspended shellfish growout systems, have had to be developed by aquaculturists themselves.

They have benefited however from the accumulated knowledge of seafarers in general and fishermen in particular, in the design and operation of mooring and buoyage systems.

More recently, when fish farmers have turned their attention to how to operate fish cages in locations further offshore where seas are rougher, the experience of the oil exploration industry has proven very valuable. Techniques have been developed in recent years for the production of fish and other aquatic products in closed recirculation systems. To make these work, aquaculturists have needed to develop a knowledge of the biological processes operating - such as how bacteria can be used to neutralise and re-cycle the nitrogenous waste products produced by growing fish - and how to engineer the systems to meet the biological requirements. Knowledge of bio-engineering from the waste treatment and water treatment industries has made contributions to the development of closed aquaculture systems and there is probably more that could be usefully transferred from the sewage treatment sector to help solve problems in fish rearing.

Modern materials have brought many benefits to aquaculture. For instance, custom made plastic joints have simplified the construction of sea cages, and made them more reliable in high stress conditions. Experiments with huge free-floating or sunken net cages operated in the open ocean were begun several decades ago, for instance in the Caspian Sea. These showed some promise, but more reliable construction materials will make the farming of fish in such structures increasingly feasible.

Modern materials and production methods have been important also, in the construction of plastic filter substrates for indoor recirculating systems. The fine detail of these has been found to make substantial differences to the efficiency of biological filters.

In shrimp farms, specially designed matting materials that stand upright on pond bottoms, with a structure that promotes the growth of the small animals and plants that the shrimp can thrive on, have recently been developed and shown to boost production.

Engineering skills are important in the design of most aquaculture facilities and good engineering can affect the efficiency and economics of production. If capital costs can be minimised while still maximising productivity and reducing risk, the farming operation will be more profitable. Aquaculturists have proven very innovative over the past fifty years, constantly developing new technologies to support their farming operations. As new production methods and species for farming are developed, the engineering solutions needed to support them will continue to evolve.

Aquaculture Facilities

Aquaculture began with man making small modifications to natural habitats so as to improve the survival and growth of target species. Some of the oldest examples are in the rearing of freshwater fish in ponds, which has been practiced for thousands of years in Asia and at least for many centuries in Europe. The simple act of placing a mesh barrier across the outlet of a small pond or lake to prevent fish from escaping can make a big improvement in the supply of food. Similarly, along coasts, some tidal lagoons can be easily turned into ponds. Closing off such naturally occurring water bodies was the start, centuries ago, of much fish and shrimp aquaculture in Asia and, in modern times, in South America.

Removing predators and improving the conditions within the pond, (for instance by providing more area of preferred water depth), supplying additional food and, later, by adding seed animals collected outside, were further steps that moved aquaculture production close to where it is today. The farming of seaweeds and molluscs (oysters, clams, mussels etc) developed similarly, as people made improvements such as providing more settlement areas for the young, or the removal of predators from the growing area.

Farmers had a natural desire to improve the productivity of their systems and, as knowledge grew, they learned to stock more animals, increase feeding and manage the exchange of water to maintain the conditions, such as adequate oxygen levels, that the animals needed to survive. The range of facilities used for aquaculture subsequently broadened for a number of reasons. Farmers encountered difficulties with more intensive rearing because of the uncontrolled influence of pond soils, local water quality, weather. Some of these problems could be resolved by rearing in ponds built of concrete, or lined with plastic, by bringing the ponds indoors under cover, or treating the water before flowing it to the culture ponds.

Secondly, as the naturally occurring ponds and lagoons all became used, prospective farmers had to take a broader approach and develop the technology and engineering to be able to use less naturally-favored sites. Net cages floating in protected coastal or inland waters were developed for fish culture, or fish were stocked in fenced areas of the sea or of large lakes. At the same time, the need for supplies of young animals (fry, seed) to stock these systems led to the development of hatchery techniques and dedicated hatchery facilities. For most species this aspect of production has proved more successful when conditions can be more closely controlled, for instance in indoor concrete and fibreglass tank systems, rather than in outdoor ponds. Often, the younger stages of aquatic animals are more sensitive than adults to physical and chemical conditions and these have to be managed within a smaller range, if production is to be successful. Thus hatchery facilities developed as a separate branch of the industry.

More recently, our knowledge has improved greatly regarding the complex interactions that occur in a rearing system, between nutrients, bacteria and the cultured organism. This and technological developments have allowed many aquatic organisms to be reared in completely closed recirculating facilities, including the farming of marine organisms at locations far from the sea. Closed systems have the added advantage of offering greater protection from the danger of disease entering from the natural environment and also of minimizing adverse effects of the production system on that external environment. Technology has also begun to open up the possibilities of growing fish in enclosures in the open ocean, something that could one day transform the nature of human food production on the planet. With 70 per cent of the earth's surface covered by water, the potential is clear. Earlier technology has restricted the cage farming of fish to sheltered coastal waters. Here the number of available sites is limited, environmental damage is more likely and conflicts exist with other users. As cages are developed that can withstand the demanding conditions of the open ocean, or which can be operated below the ocean surface, farming could move further offshore.

A small start has been made, notably in Japan, through so-called 'ranching' programmes', the farming of the sea without enclosures where fry are released in large numbers into the ocean with the aim of improving returns from the capture fishery. Sometimes artificial reef structures are created underwater as well, to increase the available habitat and natural food for the fish.

Aquaculture Systems

Aquaculture - the growing of aquatic animals and plants - covers a wide range of species and methods. Some of the simplest production systems are the small family ponds in tropical countries where carp are reared for domestic consumption. At the other end of the scale are high technology systems, such as the intensive indoor closed units used in the USA for the rearing of striped bass or the sea cages used in Chile and Europe for growing salmon and bream. Nearly half of the world's aquaculture however is of herbivorous and filter-feeding fish such as carp.

Practised in freshwater, in ponds made of earth, and often managed by a single household, production systems are simple and in many cases have changed little over many centuries. A farmer will excavate a small pond near his house; sometimes, as for instance in the flood river deltas of countries like Vietnam, using the spoil to build a raised foundation for the house itself. The pond will typically be fed by natural rainfall or groundwater, or sometimes by diverting a nearby stream or irrigation canal. Some stocking occurs naturally, but more commonly a farmer will buy young fish from a breeder and stock them in his pond, often after holding them for a time in a small floating net known as a 'hapa' to check that they are in good condition before release in the pond. Production will improve if the farmer can stock a good balance of different species to make the best use of the varied kinds of food available in the pond. Typically the fish will be fed with household or agricultural by-products, and harvested when the family needs a meal.

Stocking and harvesting is often continuous, where knowledge of the best stocking level is learned from experience and where the pond is not drained for many years. For most such farmers, fish production is a secondary activity, a useful additional source of protein to add to the supply or income from his main agricultural or commercial activities.

Outdoor freshwater fish farming is practised commercially in many countries, including in the developed nations. The density of stocking may be much higher, the control of feeding, water quality and fish health more closely monitored, but even in intensively operated systems, the key parameters that need to be monitored and balanced for success are similar to those in the traditional household pond system.

Over the last half-century, systems for the indoor rearing of many species of fish have also been developed. Holding of fish in controlled conditions indoors has been important in the development of seed production (hatcheries). Also, as knowledge has increased of nutrient cycles, bacterial action and water chemistry, it has become possible to rear fish, both for food and for ornament, in 'closed' indoor systems, where the water is recirculated. Passing the water through filters that use bacteria to naturally break down and recycle the waste products produced by the fish allows a production system to be run in almost total isolation from the outside. Such systems have been important in the control of disease and also in maintaining the stable conditions that some species need to flourish and reproduce.

A fifth of current world aquaculture production is of plants - mainly seaweeds, that for the most part are grown on the seabed or on raft or racks in shallow coastal waters and used directly for food or for the production of alginate or carageenan (agar-agar). This sector of aquaculture is another largely run by small-scale growers, mainly in Asia but in some parts of the South America, who attach seedlings of marine plants to simple structures built close to shore and constructed from natural materials and then harvest the plants once they have grown. A further fifth of world production is of molluscs such as oysters, clams and mussels.

Again the majority of production is small-scale and by coastal people who often combine fishing activities with farming. Oysters and mussels are mainly grown on structures built above the seabed - poles or racks on the shore, or ropes suspended from rafts or floating lines. The farmer's role is to supply suitable places for seed to settle or to add to natural production by bringing seed from a hatchery - and then to maintain the conditions of waterflow and freedom from predators that the shellfish need to grow. Most commercially grown molluscs feed on microscopic algae floating in the water, so the farmer does not need to providc any feed.

Fish are also grown in coastal areas, in ponds or in floating cages. As technology advances, there is the potential to develop systems for rearing fish in the open ocean, either in sturdy cages or by so called 'ranching', where young fish are released to the wild and then collected by normal fishing or by training them to respond to specially generated sounds.

The tropical coastal zones of Asia and Latin America are where most shrimp farms are found. Shrimp farming is mainly carried out in earth-bottomed ponds built on flat land close to the sea. Shrimp are raised in ponds at a range of different densities. At low densities the systems need only limited inputs of feed and fertilizers. As the density increases, more feed has to be supplied and at the higher end of the range, so-called 'intensive' farming, machines have to be put in the ponds to mix and aerate the water.

Aquaculture is practised from the cold waters of the far north and south, where fish like salmon, arctic char and sturgeon are grown in ponds, flowing raceways and cages in the sea, down through the latitudes to the tropics, where carp and tilapia flourish in freshwater and shrimp and seabass are farmed along the coasts. It ranges from production of fish in naturally occurring ponds in rural areas to the intensive culture of ornamental fish in plastic tanks in the middle of a city. It is practised by the poorest farmers in developing countries as a livelihood and to supply much needed protein for their families - and by urban sports shop owners in Europe and the U.S.A. producing baitfish for weekend anglers. Systems can range from an intensive indoor system monitored with high-tech equipment through to the simple release of baby fish to the sea - but all with the same aim, of helping boost Nature's natural productivity

Biotechnology Development

Biotechnology - any technological application that uses biological systems, living organisms, or derivatives thereof, to make or modify products or processes for specific use 1994 UNEP Convention on Biological Diversity - has a wide range of useful applications in fisheries and aquaculture. It creates opportunities, for instance, to increase growth rate in farmed species, boost the nutritional value of aquafeeds, improve fish health, help restore and protect environments, extend the range of aquatic species and improve the management and conservation of wild stocks.

Some biotechnologies are simple with a long history of application such as the fertilization of ponds to increase feed availability. Others are more advanced and take advantage of increasing knowledge of molecular biology and genetics, *e.g.* genetic engineering and DNA disease diagnosis.The field of genetic biotechnology similarly ranges from simple techniques such as hybridization, to more complex processes such as the transfer of specific genes between species to create GMOs (genetically modified organisms). Over the years, our knowledge of fish breeding requirements has improved and the ability to induce artificial breeding developed through the use of natural or synthetic hormones and/or environmental manipulations. (For example changing photoperiod or water temperature can induce some fish to spawn). These have been key factors facilitating the application of more advanced biotechnologies.

Selective breeding, the maintenance of stocks genetically improved by chromosome manipulation, line crossing, and sex reversal all depend on the controlled breeding of farmed species. These improvements in reproductive technologies have also assisted aquaculturists greatly in their efforts to domesticate aquatic animals. In addition, by making it possible to remove the natural constraints and timing of breeding, farmers are able to mate many more species at times that are most beneficial, and thus helping to ensure a steady and consistent supply of fish for consumption.

Current Status of Aquaculture Genetics

Domestication and Strain Evaluation

When wild fish are moved to aquaculture settings, a new set of selective pressures comes into play that often changes gene frequencies. Thus an organism better suited for the aquaculture environment begins to evolve. This process, termed domestication, occurs even without directed selection by the fish culturist. Domestication effects can be observed in some fish within as few as one to two generations after removal from the natural environment. In channel catfish (Ictalurus punctatus) an increased growth rate of three to six per cent per generation was observed.

The oldest domesticated strain of channel catfish (89 years), the Kansas strain, has the fastest growth rate of all strains of channel catfish. Although most domesticated strains usually perform better in the aquaculture environment than wild strains, there are exceptions, *e.g.*, wild Nile tilapia, Oreochromis niloticus, and rohu, Labeo rohita, grow better in the aquaculture environment.

The explanation for this appears related to a lack of maintenance of genetic quality and genetic degradation in domesticated strains. Poor performance of some domestic tilapia is related to poor founding (parental) lines, random genetic drift, inbreeding and introgression with slower growing species, such as O. mossambicus and slower growing strains such as Nile tilapia from Ghana. Channel catfish strains differ in growth, disease resistance, body conformation, dressing percentage, vulnerability to angling and seining, age of maturity, time of spawning, fecundity and egg size. Strains of rainbow trout, Oncorhynchus mykiss, show similar variability. Domestication of farmed shrimp (penaeids) has been relatively slow compared to that of finfish.

This can be attributed to use of wild broodstock and postlarvae, a lack of understanding of shrimp reproductive biology for domestication of the species and perceptions of low potential for genetic improvement. Current reliance on wild broodstock is risky and negates the opportunity to enhance disease resistance (as well as other production traits) through selective breeding. Efforts to domesticate broodstock are now hampered by endemic disease challenges; however, recent collaborative research between the Commonwealth Scientific and Industrial Research Organisation (CSIRO) in Australia and the shrimp culture industry has resulted in successful captive breeding of Penaeus japonicus. Economic analysis has demonstrated that domesticated broodstock are more cost-effective than wild broodstock and that reproductive performance of domesticated P. monodon can match that of wild broodstock of a similar sise.

Use of established, high-performance domestic strains is the first step in applying genetic principles to improved aquaculture mana-gement. Strain variation is also important, since there is a strain effect on other genetic enhancement approaches, such as intraspecific crossbreeding, interspecific hybridisation, sex control and genetic engineering.

Inbreeding and Maintenance of Genetic Quality

It is as important to prevent production losses due to inbreeding as it is to increase production from genetic enhancement. This applies especially to species with high fecundity, *e.g.* Indian and Chinese carps, where few broodstock are necessary to meet demands for fry and broodstock replacement. The detrimental effects of inbreeding are well documented and can result in decreases of 30 per cent or greater in growth production, survival and reproduction.

Intraspecific Crossbreeding

Intraspecific crossbreeding (crossing of different strains) may increase growth rate but heterosis (differences between offspring and parents) may not be obtained in every case. Increases of 55 per cent and 22 per cent in growth rate of channel catfish and rainbow trout crossbreeds, respectively, were achieved using this technique. Chum salmon crossbreeds, however, have shown no increase in growth rates compared with parent strains. Common carp crossbreeds generally express low levels of heterosis, however, those that exhibited positive heterosis are now the basis for carp aquaculture in Israel, Vietnam, China and Hungary.

The crossing of common carp lines in Szarvas, Hungary is a good example of the relative success of crossbreeding. During the last 35 years, more than 140 crosses have been tested.

Three were chosen, based on ~20 per cent improvement in growth rate (and other qualitative features), compared to parent and control carp lines, for culture purposes. Now approximately 80 per cent of common carp production comes from these "Szarvas" crossbreeds. Production of gynogenetic female lines and gynogenetic sex-reversed inbred male lines from common carp with the best combining ability was an important part of the Hungarian crossbreeding programme. A higher heterosis was expected from crossing inbred lines, but the growth rate of F1 crossbreeds was only 10 per cent higher than controls.

Kirpichnikov successfully produced a new strain of cold resistant carp - the "Ropsha" carp - for cold zones in northern Russia using local carps and Siberian wild carps from the River Amur. The Czech Republic also had a national common carp breeding programme and improved growth rates with crossing, "South Bohemian" x "Northern mirror" carp and the "Hungarian 15" x "Northern mirror". Likewise in Israel, over 20 years of crossing common carp strains revealed that the cross using the strain "DOR-70" and the Croatian line "Nawice" produced fast growth, and this is now one of the most popular crosses for Israeli carp production.

In Indonesia, strain development using artificial gynogenesis and sex-reversal resulted in 10 common carp inbred lines, which were later used for crossbreeding. In Vietnam, eight local varieties of common carp, along with "Hungary", "Ukraine", "Indonesia" and "Czech" strains are maintained, with significant heterosis observed in F1 generations of crossbreeds.

Double crosses among Vietnamese, Hungarian and Indonesian strains have subsequently been used for carp selection and crossbreeding programmes throughout Vietnam.

The Vietnamese x Hungarian common carp crossbreed is particularly popular, due to fast growth and high survival rates under different production conditions. Under various rice-field conditions, growth rates of different strains of Nile tilapia and their crosses showed that the crosses were superior to pure Senegal strains.

Breeding programmes are also under development in several countries for the Java or silver barb, Barbodes (formerly Puntius) gonionotus, another economically important fish species in Southeast Asia. The Bangladesh programme used three strains: "Bangladesh", "Thailand" and "Indonesian".

The growth rate of females from six crosses was 23 per cent higher than the average growth rate of the parent strains. Even higher growth rates (35 per cent improvement) were found in the three crosses using the Thailand strain as either the sire or dam. In the Vietnamese breeding programme, six different strains were used to produce a population ideally suited for culture.

Cross-breeds of different strains of European catfish, Silurus glanis, are characterised by outstanding adaptability under warmwater holding conditions and mixed diet feeding regimes.

In addition, crosses with the walking catfish, Clarias macrocephalus, have shown improved resistance to Aeromonas hydrophila infections. Studies on "domestic a" × "domestic b" channel catfish also showed greater heterotic growth rates than "domestic" × "wild" crosses (Dunham and Smitherman, 1983. The same was found in rainbow trout crosses. Strain mating incompatibilities, however, can occur and can impede fry output.

Approaches the genetics of behaviour

We describe in order of increasing complexity how a researcher might approach the genetics of a behaviour of interest. For each method, I describe its aim, the general approach, its strength and weaknesses and the kinds of traits and animals for which it is most suited.

'Low tech' approaches

The first step in approaching the genetics of a behavioural trait is to determine if the behaviour is repeatable, which involves measuring the same behaviour on the same individual several times. Repeatability is the proportion of phenotypic variance attributable to the individual, which could be caused by additive genetic variance or environmental variance with long-lasting effects. Evidence that a behaviour is repeatable indicates that it might have a heritable component and is amenable for further genetic dissection.

The repeatability of a trait sets an 'upper bound' to heritability. However, it is worth considering that non-genetic effects can also produce stable behaviour; repeated reinforcement including learning can also lead to stability. Another indication that there might be a heritable component to the behaviour is if it is consistent across contexts; in this case, the stability occurs via a correlation between individual behaviours in different contexts, or a behavioural syndrome.

Another relatively low-tech approach to the genetics of a behaviour is to ask if it differs across populations that occur in different kinds of environments. If so, then population differences in the behaviour could reflect an evolved response to differing selective pressures. Alternatively, the differences could reflect environmentally-induced changes within individuals.

A way to disentangle these two hypotheses is to rear animals from the different populations in a 'common garden', or in the same environment. If population differences in the behaviour are preserved under common environmental conditions, then the difference between populations might be genetic in origin.

A common garden experiment represents the first step towards approaching the genetics of complex traits such as behaviour, but there are at least two important factors to consider before concluding that population differences are genetic in origin. The first is that F1s reared in a common garden could still experience parental effects from their wild-caught parents.

To control for such environmental effects, it is preferable to rear the offspring of lab-born individuals and to compare the phenotype of the F2s. Second, GxE interactions should also be considered when populations do not differ in a common environment. For example, individuals from both populations might converge in their response to the common garden environment, but the extent of response might have a genetic component. A reciprocal transplant experiment might reveal genetic differences that are not apparent in a common environment.

An advantage to using population differences as a clue to which behaviours are likely to be heritable is that traits that differ across populations in different environments are probably linked to fitness. Several common garden studies on fishes have shown that population differences in behaviour are often preserved in a common garden. For example, Lahti *et al.* 2001 found evidence for a genetic basis to aggressive behaviour in brown trout by comparing different types of populations. The authors reared fish from 10 populations in a common environment. The populations can be grouped into the following types: resident (non-migratory), sea-run (migratory) or lake-run (migratory), so there were replicate populations for each type. Contrary to the expectation that resident trout are more aggressive than migratory forms, they found that sea-run populations were consistently more overtly aggressive towards opponents than the other types of populations.

There is a rich literature comparing domesticated and wild populations of salmonids, which has revealed genetic influences on several types of behaviour (reviewed in). When reared in a common garden, hatchery fish show reduced antipredator responses are more bold towards a novel object and are frequently more aggressive. However, wild Atlantic salmon (*Salmo salar*) became dominant over hatchery salmon if they were given an opportunity to establish residence, suggesting that the expression of genetic differences between hatchery and wild salmon depends on the environmental context.

Quantitative Genetic Approaches

There is a long history of examining the genetic basis of continuously distributed traits using quantitative genetic techniques, which are based on the phenotypic resemblance among relatives due to shared genes. Unlike some of the reverse methods, quantitative genetic techniques measure phenotypes, not genes, and rely on the resemblance among relatives to infer genetic effects. Quantitative genetic approaches assume that many genes of small effect influence the phenotype, thereby producing a continuous distribution (however, recent QTL studies have called this assumption into question). By measuring the phenotype on individuals of known relatedness, we can partition the phenotypic variation into environmental, genetic and gene x environment variance components.

The simplest quantitative genetic approach is to ask whether the trait of interest runs in families, which would suggest that there might be a genetic component to it. However, families often share a common environment so cross fostering experiments are useful here. Like repeatability, a demonstration of full-sib resemblance sets an upper limit to the heritability. More complicated breeding designs involve estimating the resemblance of parents and offspring using regression, or generating full and half-sib families to disentangle parental effects from genetic effects using ANOVA-based approaches, which are thoroughly described elsewhere. These approaches estimate heritabilities, genetic correlations, parental effects and non-additive genetic variance.

The most powerful quantitative genetic experiments estimate the G matrix. The G matrix refers to the multivariate matrix of genetic variances and covariances between several different traits. Although laborious to quantify, obtaining a G matrix allows us to predict the consequences of selection on any given trait, and, with a few assumptions about the stability of the G matrix through time, allows us to retrospectively analyze selection in the past. In addition to satisfying the assumption of stability required for retrospective selection analysis, it is also desirable to compare G matrices to determine whether genetic constraints are limiting. If genetic correlations can be uncoupled, then the relationship between traits might respond to selection, and the G matrix will, itself, evolve. Alternatively, genetic correlations might limit the number of different configuration of traits that are possible, which would be reflected in an invariant G matrix through time. Fortunately, there are several different techniques available to compare the structure of G matrices but there is debate about the best method.

A relatively new powerful technique known as the animal model allows us to estimate genetic variances and covariances without performing breeding experiments in the lab. Therefore this method is particularly useful for field studies. The animal model is powerful because it takes advantage of all types of resemblance between relatives to partition variance components; it uses all the available information about relatedness from a pedigree. Moreover, this method is suitable for unbalanced datasets, and can accommodate missing values.

Another approach for estimating genetic variances and covariances is to perform an artificial selection experiment, which involves selective breeding of individuals.

Selection experiments have shown that many behavioural traits can respond to selection. Artificial selection experiments are only suitable for organisms that can be kept in the lab, and which ideally have a short generation time. Selection experiments should only be undertaken when different selected lines can be replicated and when sample sizes are sufficient to control for drift and inbreeding. The pitfalls and perils of selection experiments are reviewed in.

Artificial selection experiments can be particularly insightful when they vary the environmental context in which selection occurs. For example, in a very interesting series of selection experiments on medaka (*Oryzias latipes*, Ruzzante and Doyle (1991, 1993) showed that the correlated response of aggressiveness to selection for fast growth depended on the ecological context in which the selection took place. In these experiments, Ruzzante and Doyle selected for fast growth under two different conditions: when food was clumped and could be defended, and when food was dispersed. The same limited amount of food was added to each treatment. Selection for fast growth produced a correlated response to selection on aggressiveness, but only when food was defensible. Interestingly, levels of aggressiveness *decreased* in the fast-growth lines, which they interpret as reflecting indirect selection for 'social tolerance'.

One of the strengths of a selection experiment is that it can tell us whether there are packages of traits that are linked together and respond to selection in concert. For example, selection for increased and decreased stress responsiveness, as measured by the change in circulating concentrations of cortisol in response to handling stress, produced several correlated effects on physiological and behavioural measures in rainbow trout. Relative to trout that did not release very much cortisol in response to handling stress, 'high responding' trout were more aggressive in a new environment were more active in the presence of an intruder and took longer to acclimate to a new environment. Moreover, these behavioural differences were also accompanied by changes in the brain monoaminergic systems. These results suggest that an entire suite of physiological and behavioural changes accompanied modifications to stress responsiveness. Interestingly, the packages of physiological and behavioural traits that changed together are analogous to different 'coping styles' that have been identified in other vertebrates.

Candidate Gene Approaches

Over the past few decades, there has been increasing evidence that the molecular functions of many genes are highly conserved across species. For example, studies on transgenics have revealed that genes from one species accomplish similar functions in distantly-related species. The incredible conservation of gene function in living things allows us to apply genetic information about other species to the organism of interest. This means that we can apply the genetic information gained from studies on model organisms to non-traditional models for which genomic information is not available.

Genes could become candidates based on genetic studies in other species (*i.e.* when polymorphism has already been identified and associated with behavioural variation, Fitzpatrick *et al.*, 2005), or because a physiological pathway leading to behaviour of interest is well-understood, suggesting hypotheses about which genes to examine. For example, the neuroendocrine mechanisms underlying circadian rhythms are well described for several mammalian species. Therefore, we can ask whether the expression or structure of genes along those pathways differ among individuals or among groups. Candidate gene expression approaches typically involve measuring the abundance of mRNA transcript in a particular tissue (usually brain, in the case of behaviour) using quantitative real-time PCR or Northern blots.

One of the attractions of the candidate gene approach is that detailed genomic information on the species is not necessary, therefore this technique can be applied to non-model organisms by designing degenerate primers based on other species. Another attraction of the candidate gene approach is that unlike quantitative genetic approaches, it does not require complicated breeding designs or data on the relationships among individuals – comparisons can be made across individuals of unknown relatedness because the genes are being measured directly. Therefore this approach is suitable for organisms that do not readily breed in the lab.

Summaries of using the candidate gene approach for studying behaviour are well-described elsewhere. Notable examples include work on the *for* gene and foraging behaviour, and the 'fruitless' gene and mating behaviour in Drosophila and the promoter region of the vasopressin gene and parental behaviour. The candidate gene approach for studying the genetics of fish behaviour is a relatively unexplored but promising future research direction. While the candidate gene approach is very attractive and has great appeal especially for non-model organisms, there are some drawbacks. First, candidate gene studies are biased; what if the ultimate source of genetic differences between groups lies further up or downstream of the particular candidate gene? One approach to this issue is to choose a pathway that is probably associated with the behaviour and look at gene expression at several points in the pathway. Second, candidate gene studies are purely correlative. Concluding that the gene is really associated with the behaviour requires further experimentation. Perhaps most seriously, candidate gene approaches work best when the trait is influenced by just few genes of major effect, which is probably the exception rather than the rule for behavioural traits.

Genomics

'Genomics' refers to the study of the structure, content and evolution of genomes, including the analysis of the expression and function of genes and proteins. What distinguishes genomics from other branches of genetics is that it looks at the whole genome simultaneously, rather than focusing on one gene at a time. Luckily, doing genomics does not require a full genome sequence, and there are already several good examples of applying whole-genome approaches to non-model organisms and ecologically-relevant traits, including behaviour. Genomics has been hailed as an opportunity to integrate mechanistic and evolutionary approaches to studying behaviour because information about genes provides neuroscientists and behavioural ecologists with a 'common language'.

Genes and Chromosomes

Genes are located on structures called "chromosomes," which are located in the nucleus of every cell. Although there are some exceptions, chromosomes typically occur as pairs in animals, and for practical fish culture management, fish can be considered to be diploids (chromosomes occur in pairs). Some rare species of fish are triploids (chromosomes occur in groups of three), but none are aquacultured species. Some species of fish are tetraploids (chromosomes occur in sets of four), notably the salmonids and catostomids; however, for practical breeding work these species behave as if they were diploids, so they can be considered to be diploids.

Because chromosomes occur in pairs, each gene also occurs as a pair. There are some exceptions in fish with morphologically distinct sex chromosomes, but few fish have such chromosomes; consequently, for practical fish breeding work, genes can be considered to occur in pairs. One chromosome of each pair comes from a fish's mother, while the other comes from its father; this means one gene from each pair comes from the mother, while the other comes from the father. The two chromosomes that form a pair are called "homologues."

Meiosis

Meiosis is the process during which the primary gametocytes develop into eggs or sperm. A number of important cellular changes occur during meiosis, but for a geneticist, the important ones are those that affect the genes and chromosomes. Meiosis consists of two cell divisions. The first meiotic division is called the "reduction division," and the second is called the "equational division." Prior to the reduction division, the homologues that form each chromosome pair replicate. The replicated homologues do not separate; they are joined together at the centromere, and the joined replicated homologues are called "sister chromatids".

The replication of the chromosomes is usually uneventful in that they replicate themselves perfectly. However, mistakes in the replication process occur. These mistakes are called "mutations." The mutation rate for each gene is low, ranging from one in 10,000 replications to one in 100,000 replications.

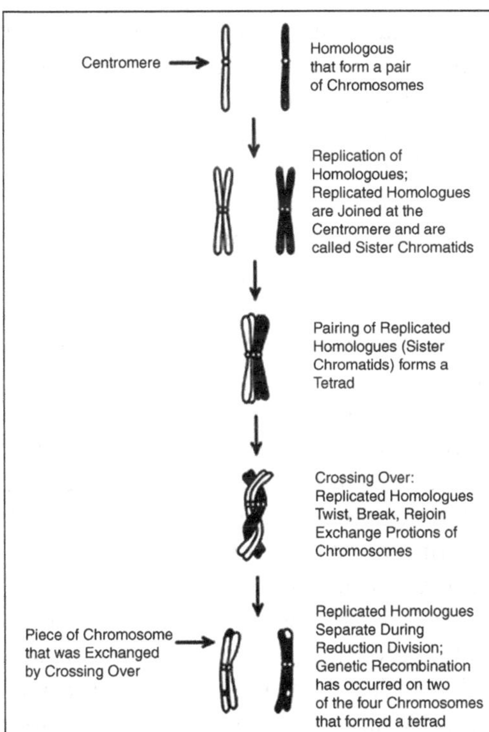

Fig. Schematic Diagram of Crossing Over, which occurs in the Initial Phase of the Reduction Division During Meiosis.

When a mutation occurs, the erroneous copy of the gene is often capable of producing an alternate version of the original phenotype (phenotypes are described in the next section). Most mutations are harmful, but some mutations create new versions of genes, and these new variants produce new or improved phenotypes that increase survival; this can lead to the evolution of new species. If replication were perfect, life would not have evolved beyond the single-celled state. Mutations also help explain some of the consequences of inbreeding.

Prior to the reduction division, the homologues that form each pair of chromosomes are replicated. This is when mutations (mistakes in replication) occur. In this figure, the homologue that came from the mother is white, while the homologue that came from the father is gray. The replicated homologues are joined at the centromere and are called sister chromatids. During the initial phase of the reduction division, the replicated homologues (sister chromatids) pair and form a bundle of four chromosomes called a tetrad. The chromosomes that form a tetrad typically twist around each other, break, and rejoin. Pieces of chromosome (and genes) can be exchanged between maternal and paternal homologues when this occurs. This process is called crossing over.

During the early phase of the reduction division, the replicated homologues that form each chromosome pair come together. Because each homologue is composed of a pair of sister chromatids, each bundle is composed of four chromosomes called a "tetrad".

The chromosomes that form a tetrad typically twist around each other and break. When the chromosomes rejoin, pieces can be exchanged between homologues. When this happens, pieces of chromosomes, along with the genes that are located along those pieces, are transferred from one homologue to the other. This process is called "crossing over". Crossing over is one of the most important biological processes, because it greatly increases genetic variability.

In inbreeding, crossing over is important because it can affect the amount of inbreeding produced during one form of chromosomal manipulation. Crossing over tends to occur at particular locations along each chromosome. Each gene has a given cross over frequency. Genes located near the centromere cross over less frequently than those further away. Gene that are located near each other on a chromosome cross over as a unit and are called "linked"; the closer they are, the more tightly linked they are. Consequently, crossing over causes actual levels of inbreeding to deviate from predicted levels during meiotic gynogenesis.

During reduction division, the chromosome number is reduced from the diploid (paired) state to the haploid (unpaired) state. The homologues that form each chromosome pair separate and are parceled into the two daughter cells (for males, the two secondary spermatocytes; for females, the secondary oocyte and first polar body). When the chromosomes divide, they are not parceled into maternal and paternal sets. The direction in which each replicated homologue (sister chromatids) of each chromosome pair goes is random and independent of that for all other chromosomes. This process is called "independent assortment". Although each secondary gametocyte contains a haploid set of chromosomes, each chromosome is duplicated.

The second meiotic division is called the "equational division". During the equational division, the replicated homologues (sister chromatids) in each secondary gametocyte separate and are parcelled into either two sperm cells or into the egg and second polar body. This produces haploid gametes; each gamete contains a single chromosome from each pair, which means that it contains a single copy of each gene.

Phenotype and Genotype

Each gene contains the chemical blueprint for the production of a protein. This protein either forms or helps produce a specific phenotype (also called "trait" or "character"), such as body colour or weight. A phenotype is the physical or chemical expression of what a gene produces; it is either observed and described (qualitative phenotype) or it is measured (quantitative phenotype). A gene can occur in one or more forms. Alternate forms of a gene are called "alleles."

In a population, a gene can have one to perhaps a dozen alleles. If there is only one allele in the population, the gene is said to be "monomorphic". If there are two or more alleles at a locus (locus = gene; plural is loci), the gene is said to be "polymorphic". Monomorphic genes are not very interesting, because there is no genetic or phenotypic variance associated with that locus. On the other hand, polymorphic genes are of great interest to a geneticist. When more than one allele exists at a locus, there is genetic variance, which produces phenotypic variance; and this can be exploited by breeding programmes.

The genetic make-up of each fish is called its "genotype". Because chromosomes occur as pairs, each gene occurs as a pair; this means the genotype is a paired entity. If the copy of the gene that a fish inherits from its father is the same as the copy that it inherits from its mother, that gene is an identical pair of alleles, and the fish is said to be "homozygous" at that locus.

If the copy of the gene that a fish inherits from its father is different from the copy that it inherits from its mother, the gene is a non-identical pair of alleles, and that fish is said to be "heterozygous" at that locus. Consequently, every gene can exist in one of two genotypic states:

homozygous (identical pair of alleles) and heterozygous (non-identical pair of alleles).

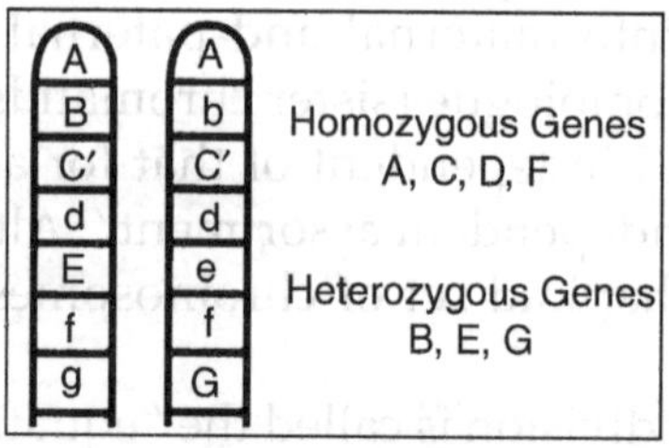

Fig. A Small Section of a Pair of Chromosomes, Showing Seven Genes. Four of the Genes Exist in the Homozygous State (A, C, D, and F), while Three Exist in the Heterozygous State (B, E, and G).

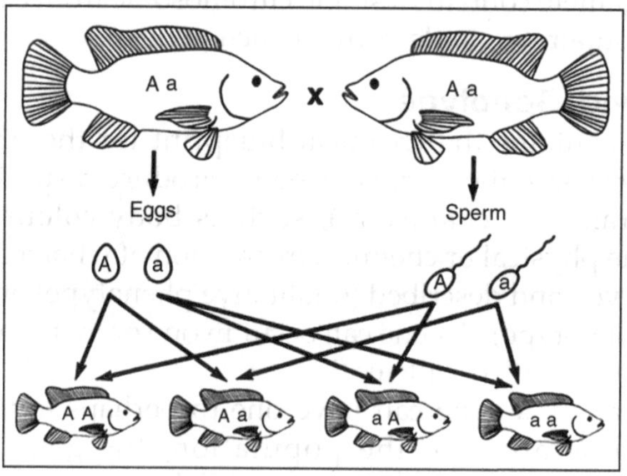

Fig. How Homozygous and Heterozygous Fish are Produced.

Progeny with *AA* and *aa* Genotypes are Homozygous; they Inherited identical alleles from both parents

Progeny with *Aa* and *aA* genotypes are heterozygous; they inherited a different allele from each parent

In this example, both parents were heterozygotes (Aa), so each produced two types of gametes: half the eggs and half the sperm carried an A allele, while the other half carried an a allele. Some progeny were homozygous (AA or aa) because they inherited the same allele from both parents. Some progeny were heterozygotes (Aa) because they inherited a different allele from each parent. The reason why it is important to make a distinction between the homozygous and heterozygous states, is they often produce different phenotypes. The number of phenotypes produced by a gene depends on the mode of gene action and the number of alleles.

A fish's genotype produces its phenotype. Consequently, if a breeder wants to conduct a breeding programme to control or alter phenotypic frequencies (using inbreeding or preventing it are both breeding programmes), he should understand how the genotype produces the phenotype.

Genetics of Qualitative Phenotypes

Breeders divide phenotypes into qualitative (descriptive) phenotypes and quantitative (measured) phenotypes. Qualitative phenotypes are those that are described, such as body colour. Because these phenotypes are descriptive, individuals segregate themselves into descriptive categories: pink vs. normally pigmented; gold vs. bronze vs. black; normal vs. deformed. In general, qualitative phenotypes are controlled by one or two genes. The alternate forms of a phenotype (pink vs. normally pigmented or normal vs. deformed) are produced by the alternate forms of a gene (the alleles).

Most qualitative phenotypes that have been described genetically in fish are controlled by a single autosomal gene with two alleles (autosomes are the pairs of chromosomes that males and females share in common; the pairs of chromosomes that are not the pair of sex chromosomes). Because of that, this discussion will concentrate on these phenotypes, especially since this type of inheritance helps explain how inbreeding works.

In general, genes express themselves in either an additive or a non-additive manner. When genes express themselves in a non-additive manner, one allele is dominant, while the other is recessive. The terms "dominant" and "recessive" carry no value judgement; the terms simply refer to the way the alleles produce their respective phenotypes.

There are two types of dominant alleles. When the dominant allele exhibits complete dominance, it always produces its phenotype, while the recessive allele can produce its phenotype only in the absence of the dominant allele. Thus, a gene that produces two phenotypes via complete dominance (it has a completely dominant allele and a recessive allele) has three genotypes, but the genotypes produce only two phenotypes:

Genotype	Phenotype
Homozygous dominant	Dominant
Heterozygous	Dominant
Homozygous recessive	Recessive

This type of gene action plays an important role in the genetics of inbreeding, and it helps explain inbreeding depression. A second type of dominance occurs when the dominant allele is incompletely dominant; *i.e.*, the dominant allele always produces its phenotype, but it is unable to completely suppress the recessive allele in the heterozygous state. When this happens, a gene with incomplete dominant gene action (it has an incompletely dominant allele and a recessive allele) has three genotypes, and each genotype produces a unique phenotype:

Genotype	Phenotype
Homozygous dominant	Dominant
Heterozygous	Near-dominant
Homozygous recessive	Recessive

The heterozygous genotype produces a phenotype that resembles but is slightly different from the dominant phenotype.

When the mode of gene action is additive, neither allele is dominant, and each allele always produces its phenotype in a unidirectional step-wise manner. This means the heterozygous phenotype is intermediate between the two homozygous phenotypes. When this occurs, a gene with additive gene action produces a unique phenotype for each genotype:

Genotype	Phenotype
Homozygous allele 1	Homozygous phenotype 1
Heterozygous	Heterozygous
Homozygous allele 2	Homozygous phenotype 2

Only one additive gene has been discovered in fish; it controls golden, palomino, and normal body colours in rainbow trout, Oncorhynchus mykiss. All other qualitative phenotypes that are produced by a single autosomal gene are controlled either by complete dominance or incomplete dominance.

Genetics of Quantitative Phenotypes

Quantitative phenotypes are those that are measured, such as length, weight, and fecundity. The important production phenotypes are quantitative ones, although some qualitative phenotypes are quite important, and can greatly increase the value of a crop. Because quantitative phenotypes are measured, each phenotype is a single category, such as length. Individuals do not get segregated into alternate phenotypic categories, such as long vs. short; instead, they are distributed along a continuum, and differences among individuals are determined by the unit of measure that is used to assess the phenotype: millimeters, grams, etc.

Because each fish's phenotype is determined by a measurement, quantitative phenotypes form what are called "continuous distributions," which are described by the population's mean and the distribution about the mean (variance and standard deviation). In a population, these phenotypes form what are called "normal" or "bell-shaped" distributions.

The reason why quantitative phenotypes exhibit continuous distributions and why individuals do not get segregated into descriptive categories is that quantitative phenotypes are far more complicated genetically than qualitative phenotypes. Each quantitative phenotype is controlled by dozens to hundreds of genes, each of which makes a small contribution to the production of the phenotype. The exact number of genes is usually never known.

Additionally, each phenotype is strongly influenced by the environment (*e.g.*, date of birth, access to food, age of mother, etc.), and the influence varies both from family to family and from individual to individual. The simultaneous action of many genes and the environmental effects creates single phenotypic categories (*e.g.*, length), in which the only way to describe an individual is to measure it.

Because each quantitative phenotype is controlled by numerous genes, as well as environmental variables, the only way to work with these phenotypes is to analyse the phenotypic variance that exists in the population and to divide the phenotypic variance into its component parts. Phenotypic variance (V_P) is the sum of the genetic variance (V_G), environmental variance (V_E), and genetic-environmental interaction variance (V_{G-E}) components:

$$V_P = V_G + V_E + V_{G-E}.$$

When conducting a breeding programme, a geneticist tries to exploit V_G. Three distinct types of genetic variance combine to make V_G, and it is important to know what they are, because different breeding programmes are needed to exploit each type. Genetic variance is the sum of additive genetic variance (V_A), dominance genetic variance (V_D), and epistatic genetic variance (V_I):

$$V_G = V_A + V_D + V_I.$$

These components of genetic variance do not refer to additive, dominance, and epistatic gene action; they refer to specific components of phenotypic variance that are produced by the entire genome, not that produced by one or two genes.

The major components are V_A and V_D. Epistatic genetic variance is usually considered to be unimportant, because it is difficult to exploit and improvements that occur by exploiting V_I plateau quickly.

Additive genetic variance is the component that is due to the additive effect of all the fish's alleles taken independently; *i.e.*, the sum of the effects that each allele makes to the production of the phenotype. Additive genetic variance is the most important component of V_P, and the percentage of V_P that is controlled by V_A is called "heritability" (h^2):

$$h^2 = V_A/V_P.$$

Additive genetic variance is the genetic component that can be exploited by selective breeding programmes. In general, phenotypes with

h^2's $\gtrsim 0.25$ can be improved efficiently by individual (mass) selection; when h^2's are $\lesssim 0.15$, family selection is needed to improve a population. When h^2 is $\lesssim 0.10$, family selection is often ineffective, although some gains can be made when inbreeding is combined with family selection.

Dominance genetic variance is the other major component of V_G. Dominance genetic variance is the component that is due to the sum of each interaction that exists between the two alleles at each locus. Because V_D is produced by the interaction of the alleles at a locus, V_D cannot be inherited from either parent. Dominance genetic variance is a function of the diploid state (a function of the paired gene), and offspring inherit alleles that exist in the haploid state from each parent. Dominance genetic variance is created at fertilization when the haploid set of chromosomes from the mother pairs with the haploid set of chromosomes from the father. When this occurs, each gene exits in the paired state, and V_D is created. Consequently, V_D is destroyed by meiosis, and it is recreated anew and indifferent combinations at fertilization.

The breeding programme needed to exploit V_D is crossbreeding. When h^2 is $\lesssim 0.15$, crossbreeding is often prescribed to exploit V_G.

Culture-based Fisheries Aquaculture

In contrast to aquaculture which involves the cultivation of aquatic life within controlled environments or the commercial production of certain aquatic species by managing the major part of their life history under strict control, culture-based fisheries increase production in natural environments by controlling a part of the life history of certain species and transplanting or releasing their seed or fry into the open waters. The juvenile fish, which are produced in hatcheries and are released into fresh-, brackish-, or marine waters, are allowed to propagate or grow on natural foods until they reach harvestable size.

Unlike aquaculture, culture-based fisheries are not limited by land or population pressures and do not have to modify or manage the culture system to approximate the natural environment since it uses the natural environment itself. Also, unlike aquaculture, harvests are uncertain and returns are more difficult to predict as the "release of fish to sea involves the number harvestable ultimately being reduced to 1-15% of the number released, depending on species and locality".

Nonetheless, culture-based fisheries have been increasingly resorted to as means of enhancing the fishery resources, replenishing natural stocks whose populations have declined through over-exploitation or environmental degradation, or simply maximizing the productivity of a water body, be it an open bay, a coastal lagoon, or a freshwater reservoir.

Their adoption or application on an expanded scale has in fact been identified as a high-priority item in the agenda for development in a number of regions in the world in the coming years. The major types of culture-based fisheries include: (i) sea/ocean ranching as practised in Japan with various marine finfishes and the Kuruma prawn, and in the Pacific Ocean and Baltic Sea with salmon; (ii) coastal lagoon farming, as exemplified by valliculture in the Mediterranean; (iii) stocking and restocking in freshwater lakes and reservoirs; and (iv) floodplain fisheries management.

Sea/Ocean Ranching

Sea ranching, in which juvenile fish are released to grow unprotected on natural foods in marine waters where they are harvested at marketable size, takes advantage of the sea as a large aquatic pasture. While it is practised in slightly differing forms around the world, in its distinctive form, young fish are released to the sea through ponds, enclosures or channels, and allowed to grow until they are ready for the market.

Salmon

Sea ranching of salmon is widely practised in countries such as the United States (mainly Alaska, Washington and Oregon), Canada, Japan, USSR, New Zealand, Iceland, and the Baltic countries.

The most widely used species for ocean ranching is the Pacific salmon (Oncorhynchus sp.) because of the success of its artificial spawning coupled with the rapid growth and the high degree of its homing precision that makes it amenable to harvesting at maturity.

The practice of salmon ranching was initially carried out more than 100 years ago in the Nordic countries where salmon smolts were released into the Baltic Sea area.

Hatchery production of salmon in the United States dates back to 1872 when the US Fish Commission established its first hatchery on the McCloud River in California. This was followed by the first private salmon hatchery on the Rogue River, Oregon in 1877. The technology for artificial propagation of salmon was introduced to Japan in 1877 and to western Canada in 1884.

In Japan, while salmon hatching and release have been carried out for nearly 100 years, the results were not considered very successful. It was only with the new method of intermediate feeding and timely release of salmon fry, starting in 1961, that significant advances in salmon production have been achieved. Most Japanese hatcheries use gravel-lined raceways to incubate larval salmon (alevins) and practise short-term feeding of newly-emerged chum fry on artificial diets.

Pink and chum salmon are raised in equal numbers in Soviet hatcheries using technology similar to that of Japan. In the Pacific Northwest, chinook and coho salmon are the predominant species used in hatcheries where raceways and ponds for feeding juvenile fish are prominently featured.

In Alaska, floating raceways in protected saltwater bays had to be innovated to suit Alaskan conditions. Canada's salmon farming industry has grown tremendously in the last five years as a result of its successful hatchery operations.

Canadian hatchery workers have pioneered development and application of spawning and egg incubation channels which provide favourable environmental conditions for eggs and alevins. These techniques have been applied most successfully to sockeye, pink, and chum salmon.

The spawning channels produce fry with a capacity for rapid growth and high survival because they simulate high-quality natural spawning beds. Incubation channels, shallow raceways with a silt-free gravel bed, enable the alevins from artificially fertilized and hatched eggs to complete their development in a natural substrate.

Salmon fry that need to be raised in low-salinity or fresh water before release into marine waters (*e.g.*, sockeye, coho, and chinook) have to be transferred from the incubation system to feed-lot systems consisting of tanks, raceways, ponds, floating pens, or a combination or modification of these. The newly emerged fry are given frequent feedings of high-protein diets containing vitamin supplements and certain fats.

Other Species

Aside from the widely popular Pacific salmon, other marine species have been used for ocean ranching in Japan. The Japanese National Culture-Based Fishery Project, which was started in 1953, has succeeded in developing techniques for mass seed production and subsequent release of juveniles of commercially valuable fish/shellfish, such as the Kuruma prawn, red sea bream, blue crabs, sole, flounder, and yellowtail, into the Seto Inland Sea.

The Project has now reached the stage where the fishermen themselves carry out the release of fry and manage the propagation work. The Project, based on its earlier successes, has been expanded on a national scale with the construction of a large number of Prefectural Centers for Culture-Based Fisheries all over Japan. The Ezo abalone, Haliotis discus Hanai, is one of the most important species for ocean ranching in Japan, with about 1.15 million hatchery-produced juveniles released in 1980 in rocky areas of the Sea of Japan around Hokkaido. Likewise, the giant Ezo scallop,

Patinopecten yessoensis, is propagated in the Okhotsk Sea, the Nemuro Channel, and in other areas of the Hokkaido coast.

Coastal Lagoon Farming

Extensive lagoon farming is a traditional activity in various European and Mediterranean countries with a production of about 10 000 t/y. Among the Mediterranean countries, Italy has the largest areas of brackish water that are exploited for lagoon farming of commercially valuable species like Anguilla anguilla, Mugil cephalus, Liza aurata, Liza saliens, Liza ramada, Chelon labrosus, Dicentrarchus labrax, and Sparus aurata.

Restocking of the valli/coastal lagoons is carried out by anadromous, natural, and annual migration and by artificial restocking. Catches from the valli primarily depend on the quantity of fingerlings which entered the lagoons some years before and on the quantity of fingerlings stocked.

Valliculture

Valliculture, as the practice is called in view of its being fish culture-based, is one of the most ancient forms of aquaculture in the Mediterranean region. Its origins date back to the first rudimental fish pond and fattening systems used along the Adriatic and Tyrrhenian coasts. This technique was developed by the upper Adriatic populations to exploit the seasonal migrations of some fish species from the sea into the lagoon and delta areas which were more suitable for their growth. The fish returned to the sea because of altered environmental conditions (temperature) of the sea or for reproduction.

To exploit these periodic movements, large brackish areas were enclosed to prevent the fish returning to the sea and complex permanent capture systems, fish barriers, were developed consisting of barriers in the channels communicating with the sea to catch the adults. Later, from the simple ponding of fry freely entering the lagoon from the sea, came a man-made seeding of fry fished elsewhere and introduced into the basins to be reared for a few years.

The successful management of coastal lagoons using hatchery-bred fry of important species is thought to hold considerable promise world-wide for expansion in the future.

Brush-park Fisheries

Brush-park fisheries are a traditional form of low-technology aquaculture which is practised in coastal lagoons and brackish waters in many areas of the world. The brush parks are constructed in a variety of forms and sizes, but basically a brush park consists of an inner core, or concentric circles, of densely packed tree branches or other material surrounded by an outer, more substantial wooden framework which is fished periodically, usually by encirclement.

In coastal lagoons the most sophisticated forms of brush-park fisheries occur in Benin. There are several basic types of brush-park fisheries in Benin where they are known collectively as "acadjas". They vary from each other in configuration, construction and fishing characteristics. The acadjas are constructed of tree branches with harder woods forming the peripheral structure and with soft wood branches with many ramifications forming the interior of the acadja. Acadjas are preferentially placed in shallow, quiet waters of no more than 1.5 m depth.

Fishing is accomplished by surrounding the smaller types of acadjas with a net after which all of the branches are removed and the net is then pursed. For the larger brush parks the surrounding net is fished in a step-wise process by gradually moving the net inwards as branches are removed until the fish are concentrated in small areas and can thus be removed with traps, baskets, and hand-nets. After fishing, the used branches are replaced and new ones added as necessary.

Elsewhere in Africa, brush-park fisheries are found in coastal lagoons in Nigeria, Côte d'Ivoire, Ghana, Togo, and Madagascar. Elsewhere in the world, brush-park fisheries are found in Negombo lagoon in Sri Lanka and have been introduced in some Mexican lagoons.

Stocking of Inland Waters

Among the most advocated and widely used practices for the management of the fisheries resources of lakes and reservoirs throughout the world is the stocking and recapture technique.

Species stocked are either indigenous or exotic and either herbivorous, carnivorous, or omnivorous. In the developing countries of the tropics, where the production of more fish for food is the goal of any fisheries activity, high-yielding herbivores, detritivores and plankton feeders (like tilapia and carp), are commonly stocked in lakes and reservoirs. This is in direct contrast with the developed countries in temperate zones where carnivores like trout and bass are stocked in lakes and reservoirs to support the recreational or sport fishery. The major lakes and reservoirs in ASEAN countries, for example, have been stocked with hatchery-produced fingerlings of a wide variety of species including tilapia, various Chinese and Indian carps, milkfish, mullet and catfish.

In China, around 65%, or 1.4 million ha, of reservoir area out of a total of two million ha are farmed and managed like fish ponds. They are stocked with fingerlings and following a period of growth, mature fish are harvested. The species used are the common carp, grass carp, silver carp, bighead carp, mud carp, and Wuchang bream (Megalobrama amblycephala) in varying densities and species ratios. Reservoir fisheries in China are almost entirely dependent on the regular release of large quantities of fingerlings because the Chinese carp do not spawn outside their area of origin. In many instances, therefore, stocked fish make up over 90% of fish catches from reservoirs.

In Israel, many water storage reservoirs (10-40 ha area, 5-7 m depth) originally constructed for irrigation, are now stocked with a range of species including tilapia, mullet and carp, usually grown in intensively managed polyculture systems. Sri Lanka's large perennial reservoirs and small seasonal tanks constitute the country's main freshwater resource. These are regularly stocked with fingerlings of the different carp species (Indian, Chinese, and common) which are produced at Government-run hatcheries spread all over the country. Since the Indian and Chinese carps do not spawn in the reservoirs/tanks, they have to be stocked regularly to sustain the fisheries.

There is thus large pressure on the State to accelerate its seed production programme to keep up with the large demand for fingerlings. It is interesting to note that unlike the Philippines and Thailand, for example, where fish seed production is largely a private sector undertaking, the Sri Lankan Ministry of Fisheries is primarily responsible for the production of carp fingerlings for distribution among the various reservoirs and seasonal tanks.

In Cuba, most of the freshwater fish is produced from seeding of the country's small to medium size reservoirs; in Colombia, some dams are stocked and restocked with mullets.

African small reservoirs are stocked mainly with tilapias and Clarias. In Zambia the most important species for stocking reservoirs are Tilapia melanopleura, T. macrochir, T. andersonii, and Haplochromis spp. In Zimbabwe the fish most commonly used for dam stocking are Oreochromis mossambicus, O. macrochir, T. rendalli, and several exotics like black bass and rainbow trout.

Aquaculture and Marine Biotechnology

Diagnostics and vaccine development in aquaculture are a major priority apart from development of suitable cell culture systems, immunostimulants, anti-bacterial and anti-cancerous agents. Studies have been supported on isolation of novel enzymes from deep-sea fungi, molecular cloning and gene expression for alkaline protease and molecular epidemiology of white spot shrimp disease in brackish water system.

DNA vaccines developed against Aeromonas are being studied for the efficacy and challenge experiments. Immunodot-based diagnostic kit is being used for detection of white spot syndrome virus (WSSV) in shrimp and prawn. An ELISA-based kit has been developed for detection of IgM of Indian major carps against Aeromonas hydrophila infection. An anti-WSSV formulation was made as a prophylactic measure in shrimp health management. Biosurfactants production was standardized from marine wastes employing the acinetobacter strains.

New source of molecular fluorescent dyes has been attempted from sea cucumber tissues for bioactive compounds and use in biomedical applications. Bioactive properties of ink gland of cuttlefish have been studied for antibacterial, antiviral and anticancer agents. Cell culture attempted on brackish water species, seabass, could lead to establish stable species-specific cell lines useful in virus isolation.

The growth-promoting factor studied on development of embryos of Indian horseshoe crab showed significant enlargement of organ development. Development of culture technology in non-conventional species like seaweeds and product and process development of commercial importance was also the part of the programme. A protocol on better aquaculture management practices was developed through recirculatory and raceway system. In service training programme in molecular biology for fisheries scientists was received greater attention in the programme funded at CIFT, CIFA, CIFE. Two programme supports on Aquaculture/ Marine Biotechnology have been supported.

Diagnostics and Vaccines Development

At College of Fisheries, Mangalore, work was carried out on the cloning and expression of virulence genes of Aeromonas hydrophila, as an important fish pathogen and the attempts were made to isolate the proteins and study the immunogenicity. At Centre for Biotechnology, Jawaharlal Nehru University, New Delhi, recombinant porin has been expressed and purified in large quantity for maintaining the consistency of the porin vaccine. The efficacy study and challenge experiments are being undertaken on immunization of mice and fish with porin DNA vaccine. Studies were undertaken on pathogenecity of the Epizootic Ulcerative Syndrome (EUS) in fish at Viswa Bharti University, Santiniketan; North Bengal University, Darjeeling and Center for Cellular and Molecular Biology (CCMB), Hyderabad.

A project was undertaken at College of Fisheries, Mangalore on expansion of panel of monoclonal antibodies and standardization of immunodot test kit for detection of white spot virus of shrimp. A simple monoclonal antibody (MAb)-based Immunodot kit for detection of WSV of shrimp at the farmers level has been developed, which is a cheap and effective alternative for expensive and complex DNA based tests. Development of DNA– and protein-based diagnostic methods of white spot disease of shrimp, their field evaluation and monitoring were undertaken at Anna University. The purified viral protein was used as diagnostic assay kit and transferred to Posideon Aquatech Limited, Chennai for commercialization. An ELISA test has been developed using rabbit antisera against carp IgM for the detection of IgM in IMC. Studies were undertaken on development of anti-WSSV formulation against anti-white spot syndrome virus compound as a prophylactic measure in shrimp health management at CAH Abdul Hakeem College, Melvisharam. A process to purify the compound responsible for inactivating the WSSV was optimized.

Cell Culture in Aquaculture

Development and characterization of cell lines from selected finfish and shell fish was undertaken at CIFE, Mumbai and a cell culture system was developed from the fish Lates calcarifer (Sea bass). At CUSAT, the cell lines developed from eyestalk sub-cultured and made available to develop such cell lines as and when required. The cells shown differentiation in to a proliferating bunch of cells and subsequently got detached as a lump of cell aggregates. The development of cell lines from seabass, Lates calcarifer was attempted at CAH Abdul Hakeem College, Melvisharam and CUSAT. The kidney cells shown fibroblastic and formed a monolayer. The cells were were deposited at NCCS.

Growth Promoting Factor(s)/Signal Transduction

A collaborative project was undertaken on identification, isolation and characterization of growth promoting and differentiation factor(s) from perivitelline fluid of developing embryos of the Indian horseshoe crab at Agharkar Research Institute, Pune, National Centre for Cell Science (NCCS), Pune and National Institute of Oceanography (NIO), Goa. Active fraction identified shown significant enlargement and enhanced compartmentalization of the heart in cardiac development and promoting activity. In another project on large-scale cultivation of amocbocytes in vitro from Indian horseshoe crab implemented by NCCS and NIO, regeneration of gill lamellae of the Indian horseshoe crab was successful in amoebocyte production.

A molecular signal intervention mechanism was developed as an alternative to eyestalk ablation at the Departments of Zoology and Biochemistry, University of Madras. A specific inhibitor that intervenes in this pathway has been shown to reduce the molt cycle of the prawn and reduces the turnover time and rearing cost and maximizes the economic returns. By decreasing the molt cycle duration, it can accelerate the growth rate of the prawn resulting in early maturation.

Feed Supplementation for Prawn

Studies on feed supplementation of highly unsaturated fatty acids (HUFAs) and immunostimulants for disease resistance in freshwater prawn was carried out at the Department of Aquaculture, College of Fisheries, Mangalore. HUFAs namely Eicosapentaenoic acid and Docosahexaenoic acid are important for larval growth. The supplementation of PUFA incorporated feed could enhance growth, survival and disease resistance in freshwater prawn.

Raceway based Shrimp Farming

A project on prototyping of raceway based third generation shrimp production technology was implemented at the Fisheries Biotechnology Centre, Fisheries College and Research Institute, Tuticorin. The protocol has been developed to increase survival of post-larvae. This in-door shrimp farming facility will be eco-friendly, economically viable and technically useful for large-scale adoption in aquafarming systems in India.

Artemia Production

The work on development of Artemia franciscana culture in the solar salt ponds and processing of Artemia cyst and biomass was undertaken at Centre for Marine Science and Technology, Manonmaniam Sundaranar University, Rajakamangalam, Tamil Nadu. Artemia production was initiated in condenser ponds and inoculated in culture ponds and the efforts are being made to utilize the abandoned area of the saltworks for Artemia cultivation.

Product/Process Development

Studies were undertaken at NIO, Goa to isolate, purify & characterize new natural fluorescent dyes and fluorescent molecules from tissues of sea cucumber Holothuria scabra and to test them as bioactivities compounds for therapeutic and industrial applications. At Kerala Agriculture University, Kochi, attempts were made to isolate bioactive substances such as antibacterial, antiviral and anticancer agents from the accessory nidamental gland and ink of cuttlefish (Sepia pharaonis ehrenberg). An antitumour peptidoglycan was isolated from cuttlefish ink and the crude preparation showed antitumour activity against mice following intraperitonial administration. The peptidoglycan was made up of protein and uronic acid rich with polysaccharide fraction and was found to be made up of five amino acids.

The screening of marine Acinetobacter genospecies for production of biosurfactant/s, purification and antimicrobial activity was undertaken at Pune University. Untapped natural sources were screened for novel biosurfactant/s and bioemulsifier-producing microorganisms using Acinetobacter strains, which showed emulsification of toluene, diesel, hexadecane and crude oil, haemolysis of human RBC's and anionic biosurfactant/s production.

Novel Enzymes

At the department of biotechnology-CUSAT, the work on molecular cloning was undertaken to develop a technology for production of alkaline protease enzyme using marine fungus, Engyodontium alba. The optimal conditions were studied for enzyme production using solid-state fermentation of wheat bran as the solid substrate. The enzyme was found relatively stable for one year. At NIO, Goa, fungi from deep-sea sediments were explored as new sources for protease. Fungal growth and protease production at low temperature and elevated hydrostatic pressure was studied to understand their adaptations to extremophilic conditions.

Inland Water Fisheries

Inland fisheries production comes mainly from natural bodies of water, the rivers and canals, the beels and haors, and the floodplains. Their importance in terms of nutrition and employment is immense.

The rivers and canals, the beels and haors

The major river systems of the country are distributed over an area of 1 million ha (or 22,000 km long) and provide a base for artisanal and subsistence fishery activities producing around 180,000 tons of fish. Over the years production has been decreasing because of increased fishing pressure and a decrease in the extend of floodplain habitats because of the construction of flood control, drainage and irrigation systems, and the consequent obstruction of fry and fingerlings from rivers. Further declines of inland fish production are anticipated when all the planned water control projects are completed. Attempts have been made to counteract declining production in rivers by improving resources management, establishing sanctuaries and enforcing regulations but DOF has difficulties to implement these, due to lack of financial and trained man-power resources.

The economic loss caused by FCDI/FCD interventions, however has been offset by increasing harvest in inland aquaculture, marine catches and shrimp culture exports. However, the decline in the inland capture fisheries has significant nutritional consequences for many Bangladeshi, since capture fisheries are a major open-access resources for the poorer segments of the population and often the only source of animal protein, essential minerals and vitamins.

There is limited scope for geographically expanding the total area of inland fisheries and this area is expected to continue declining because of FCDI/FCD development. However, there is substantial scope for increasing fish yields and the area under aquaculture.

Floodplain Fishery

The term floodplain is generally applied to low-lying areas which are seasonally flooded by rivers. The flood plains cover a highly variable area which may be estimated at approximately 2.9 million ha during the flood season. The nature of the capture fishery in the flood plains is bound to the pattern of flooding during the monsoon season, at which time a major expansion of fish biomass takes place.

The ability of flood plains and river systems to sustain themselves depends on the extensive system of interconnected areas of habitat which provide for reproduction, migration, feeding and growth. Almost all rural dwellers are engaged in part-time fishing in the flood season when the agricultural work load is low. Capture methods are generally rudimentary, and stock management, law enforcement and conservation practices aimed at sustaining production in the floodplains are difficult to apply, hence not implemented.

Floodplains are the food-rich breeding nursery and growth areas. Many of the fish species of economic importance migrate with rising and receding floods between rivers and the flood plains. At low water levels fish retreat to rivers.

Therefore, any development in the floodplains would also affect the fish population in the river system. The flood plains and the rivers together form an important economic entity. The decrease in the floodplain area as a result of expanding flood control, drainage and irrigation schemes in the recent years has resulted in fish production, below its potential, and blocking the exchange between rivers and floodplains. This process of expanding flood control and irrigation works, may be necessary from the macro-economic policy point of view as it generates significant economic gains for the rural populations and agriculture, but utmost efforts must be made to mitigate their negative effects to fish production through appropriate designs.

Not only annual yields, but also both Hilsa and major carp stocks in floodplain have declined possibly because of FCDI/FCD development projects, embankments, sedimentation and overfishing. This can be offset by (a) the rehabilitation of Hilsa stocks and (b) the adoption of cultured-based fisheries technology and management which involves continuously stocking floodplain with appropriate carp species at optimum stocking ratios and densities. The World Bank has been implementing a large scale stocking programme to increase fish stocks in floodplain under the Third Fisheries Project (TFP) since 1991. The World Bank indicated that floodplain fishery production could reach between 1,090 and 1,410 thousand tons in the year 2010. This implies an annual growth rate of 4.5% to 5.5%. The potential for the increased production in the floodplain may be considerable, but realizing it will require sustained policies and action to facilitate the projected rate of growth.

Seaweed Cultivation

At CSMCRI, Bhavnagar, the project was undertaken on scale-up cultivation and processing of phycocolloid seaweeds (Gelidiella acerosa and Eucheuma). Multilocational trials undertaken to validate different culture methods for large-scale cultivation of Eucheuma proved that the raft cultivation method consistently yields better harvest. This method has been successfully demonstrated in the Gulf of Mannar, Tamil Nadu coast as well as at Okha, Gujarat coast. High quality agar was extracted from Gelidiella acerosa and Gracilaria sp. The Gracilaria agar had very high gel strength (1200 g cm-2) with low gelling temperature. This agar will be useful in bacteriological and molecular biology applications.

Bioreactor for Nitrifying Water for Prawn Hatcheries

The nitrifying bioreactor technology developed at CUSAT paves the way for transforming the open prawn/shrimp larval rearing systems to closed re-circulation systems.

The reactor, on properly activating and deploying in the hatchery system, can take care of the entire water quality problems. Apart from nitrification this duel capability lowers down the biological oxygen demand, total bacterial population along with stabilization of Vibrio population.

In-service Training Programme in Molecular Biology

The training programme for fisheries scientists was undertaken at three institutes namely Central Institute of Freshwater Aquaculture (CIFA), Bhubaneshwar, Central Institute Fisheries Technology (CIFT), Cochin and Central Institute of Fisheries Education (CIFE), Mumbai to train in service fisheries scientists on aspects of modern biology, health management and conservation of resources etc.

Trainees are being given both theoretical hands-on training and independent project work on molecular biology aspects relevant to fisheries topics. After successful completion of the training programme, the trainees have started working in the area of molecular biology in marine sciences at their institutes.

New Programme Support

The programme on aquaculture and marine biotechnology has been identified for support after a detailed discussion under the Task Force on Aquaculture and Marine Biotechnology.

Two proposals from the UAS, College of Fisheries, Mangalore and the Cochin University of Science and Technology (CUSAT), Cochin have been supported for attention on brood stock and larval health management of black tiger shrimp and freshwater prawn and screening for potential anticancer agents from marine Cnidarians, bacteria & fungi.

Genotypic characterization and sequencing of genes of interest of marine bacteria, fish and shellfish immunology and screening for useful natural products from microorganisms, capacity building through hands-on training programmes for scholars and technicians on molecular techniques for detection of pathogens and health management in shrimp hatchery system, microbial taxonomy, crustacean immunology and screening and characterization of marine microbes for useful products etc. are also the part of the programme.

Shrimp Genomics

Shrimp (Penaeus monodon) is a very important species for the country from the commercial viewpoint. Unfortunately, the species is not domesticated and the harvests are mainly from the wild. The reproductive biology is complex and the disease burden very high, especially viral diseases. There are populations, which are naturally resistant to viral attack and could be tapped to our advantage.

The network programme on shrimp genomics has been developed covering the aspects *viz.* functional genomics in disease resistance of Penaeus monodon, gene regulation of maturation and breeding in P. monodon, microsatellites and physical mapping of P. monodon and genomics of Penaeus indicus for comparative genomics.

The Fisheries Sector

Bangladesh is endowed with vast marine, coastal and inland waters having great fisheries potential: a 480 km long coastal line and approximately one million hectares of territorial waters extending 19 km up to the sea. The nation's economic zone extends 320 km out into the sea from the coast line. Total fish production in 1989–90 amounted to 847,000 tons, of which 72% was derived from inland fisheries and 28% from marine fisheries. Of inland fisheries, 69% was from open water capture fisheries and 31% from closed water culture fisheries (ponds, beels, baors and brackish water aquaculture). Clearly inland open water capture fisheries are the most important part of the fisheries sector, accounting for 49% of total fish production. So even small improvements in the average yield of inland open water capture fisheries could significantly affect national fish production and consumption.

Levels of genetic differentiation in aquatic organisms

Many studies have been carried out to estimate the amount of genetic subdivision within aquatic species. Levels of population substructure in various groups. On average, these levels of G_{ST} (which is equivalent to F_{ST}) are not too far from 0.2, suggesting that, in general, subdivision of populations within species is close to that which allows genetic drift to change allele frequencies in local populations in the face of gene flow by migration.

However, these average values hide a wide range of variation in population structure between species. For example, let us take two contrasting cases – the salmon, *Salmo salar*, and the mussel, *Mytilus edulis*. The salmon has a life history involving extensive juvenile and adult migration, with mature adults returning to their natal rivers to spawn. Thus, although salmon can be found over a very wide area, they are genetically very much restricted to local populations in their own rivers as this is where the transmission of genes between generations occurs. There are strong barriers to gene flow because very few fish return by mistake to non-natal rivers. In addition, each local population consists of relatively few individuals and the population can be sometimes even further subdivided into year classes that return to spawn in the same year as one another.

Values of F_{ST} for salmon, calculated from extensive allozyme data, are in the region of 0.4 and indicate that, as expected, there is very little gene flow ($N_e m$ = 0.38) and strong genetic differentiation into local populations in this species.

On the other hand, the mussel is a sedentary organism and adults never actively move more than a very short distance. As with salmon, mussels release eggs and spermatozoa into the water, but, in contrast to salmon, mussel larvae are small and planktonic and are at the mercy of currents that can disperse them over great distances. Larval life, from egg to metamorphosis, in mussels lasts around 4 weeks and can be further extended in two ways.

Firstly, if a suitable habitat for settlement is not discovered, metamorphosis can be delayed for more than a week, allowing further dispersal. Secondly, once metamorphosed, these very young mussels (spat) can detach, secrete a very thin byssal thread and be transported in currents by a process known as byssal drifting, which is analogous to the aerial dispersal of spiders by gossamer threads. Typically, F_{ST} values from allozyme studies in mussels are very low (<0.001) and clearly reflect extensive gene flow ($N_e m > 250$) and a lack of any population substructure.

It is also interesting to compare genetic differentiation in marine pelagic fish such as Atlantic herring (*Clupeus harengus*) with that in anadromous fish such as salmon. The herring has a very large effective population size because all fish return to the same region to spawn and there are therefore few, if any, barriers to gene flow. Average F_{ST} for herring is 0.01 and $N_e m$ is 24.8.

Therefore, the conclusion is that where the effective population size and/or migration is large, gene flow will tend to dominate over random genetic drift, there will be little differentiation and F_{ST} will be close to 0. Where $N_e m$ is small, random genetic drift will tend to dominate over gene flow, allele frequencies will differ strongly between populations and F_{ST} will be large.

Although information about population differentiation is very valuable to fishery managers, it is important to recognise that a *lack of evidence* for local genetic populations does not always mean a *lack of substructure.* If populations are genuinely genetically isolated from one another, then we would expect random genetic drift to be acting to change allele frequencies at all polymorphic loci. However, for reasons of chance, not all loci will show different allele frequencies in different populations. If the loci screened in limited surveys happen to be these invariant loci, then the true variation remains hidden. It is therefore important to examine sufficient loci to be sure that an apparent lack of genetic differentiation can be relied upon.

Potential problems with allozymes and coding markers

There is a risk that allozyme markers could fail to detect genetic isolation because, as enzymes, they can be vulnerable to selection operating at the biochemical level, either directly on them or on the biochemical pathways within which they operate. This can be also the case for SNP markers located in exons or any marker closely linked to another one that is subject to selection.

If the selection is such that it is in the same direction in two genetically isolated populations, that is an allele is at high frequency because the individuals carrying it are favoured in both populations, then the allozyme data will be interpreted as indicating that the populations are not genetically different. On the other hand, two populations with habitat differences but which regularly exchange genes via larval flow may have different allele frequencies at a locus which is under selection where one allele is favoured in one population but is deleterious in the other.

Differential mortalities in the two populations will cause allele frequency differences between samples taken from the two populations and this will give a false impression about the degree of reproductive separation of the two. This is why the use of several markers and, even better, different types of markers is recommended.

Here is a good example of how different makers can deliver a different message. The American oyster, *Crassostrea virginica*, occurs in estuarine habitats throughout the east coast of the USA, from Maine in the north to the Gulf of Mexico in the south.

It has a larval phase lasting several weeks and might be expected to be relatively genetically homogeneous as a result of this. Allozyme studies detected no significant genetic differentiation into local populations anywhere in the species' range. However, mtDNA markers and some neutral, non-coding nuclear DNA markers demonstrated a very distinct division into a southern group and a northern group of populations separated at a zone of demarcation near to Cape Canaveral on the Florida Atlantic coast.

This evidence makes it clear that there is practically no gene flow between the southern and northern regions, probably owing to the characteristics of a major oceanic current – the Gulf Stream – which emerges from the Gulf of Mexico around the tip of Florida.

On the basis that evolution is faster in the mtDNA molecule than at protein coding loci, it could be suggested that, although differences in the mtDNA molecule have evolved, insufficient time has elapsed since the divergence of the two groups of oysters for allozymes to reflect the division.

However, the non-allozymic single copy nuclear DNA markers also show this divergence, and the most convincing conclusion is that selection at the studied allozyme loci is maintaining the same alleles at similar frequencies on both sides of the break point. It is notable that several other species such as the horseshoe crab (*Limulus polyphenus*), the black sea bass (*Centropristis striata*) and the seaside sparrow (*Ammodramus maritimus*) also exhibit a fundamental mtDNA divergence in the same region, and this is congruent with the biogeography of the region.

What then are we saying here? Is it really the case that allozyme data, and more generally coding markers, should not be trusted at all? This is a vexed question, but the answer is that ideally, a large number of loci of different types should be used to provide a reliable picture of gene flow and population differentiation. Fortunately, selectively neutral markers such as mtDNA, microsatellites and SNPs, generally more rapidly evolving than allozymes, are now the tools of choice for studies of population differentiation.

mtDNA Variation

An example of the use of mtDNA variation to detect population structure. The anadromous fish *Coregonus artedii*, the cisco, is important in the local fisheries of the river systems of the James and Hudson Bays in Canada. Al-lozymes were just not variable enough to be able to reveal differences between local river populations; therefore, a mtDNA approach was used by Bernatchez and Dodson to investigate population structure. The frequencies of three mtDNA clonal groups (A, B and C) of cisco from rivers feeding into the Hudson and James Bays.

The clonal groups are not individual mtDNA haplotypes, but come from three clear clusters on a dendrogram of the 19 haplotypes detected from the 141 fish sampled in the study. Different mtDNA haplotypes can be compared to estimate their relatedness to one another and this is what provides clusters of more closely related haplotypes. The genetic pattern revealed by a detailed mtDNA analysis of the different haplotypes within the clonal groups suggests that the cisco of this region, which was covered by ice in the last glaciation, were actually derived from two quite separate glacial refugia (clonal groups A and B) and have recolonised the area by different routes.

Since recolonisation, there has been some merging between the two types but this has not obscured the historical geographic pattern. The source of clonal group C which is strongly diverged from both the A and B clonal groups is not clear – it might even represent introgressed mtDNA from a closely related hybridising species. In this example, we see how mtDNA analysis can not only tell us about the current population structure of a species, but also can point to historical processes which have influenced the pattern observed in the present day. In human genetics, this valuable characteristic of mtDNA has been used to provide evidence for the 'out of Africa' theory, since it appears that we are all descended from a single 'mitochondrial Eve' lineage that inhabited Africa some 140 000–290 000 years ago.

Microsatellite Variation

The rate of change at microsatellite loci is much faster than for allozyme loci or for mtDNA, and therefore, microsatellite variation is a more sensitive measure of weak genetic differences between populations. How much genetic differentiation can be resolved in salmon (*Salmo salar*) populations by microsatellites than with allozymes. The top dendrogram is based on an allele frequency analysis of 38 allozyme loci in populations from 31 rivers that drain into three main regions – the west Atlantic, the east Atlantic and the Baltic Sea. In the lower dendrogram, equivalent results based on four microsatellite loci are displayed.

The microsatellite genetic distances between river populations in the western Atlantic (*D* ranges from 0.06 to 0.11) are actually greater than the allozyme genetic distances when comparing salmon populations from both sides of the Atlantic Ocean (*D* = 0.04). So the power of microsatellites to identify population differentiation is very high. However, there are two caveats that must be added here. Firstly, the resolving power of loci with many alleles, such as microsatellites, is highly dependent on sample size. A small sample of, say, ten individuals scored at a locus with 25 alleles will inevitably only be sampling a proportion of those alleles. Comparing this with another small sample from a different population that also has the same alleles at similar frequencies could show quite large genetic differences based on chance alone. Some alleles sampled from one population will be absent in the sample from the other population, leading to the mistaken conclusion that the two populations are very different. Secondly, Nei's genetic distance measure *(D)* was devised on the basis of estimated point mutation rates at allozyme loci (around 10^{-7} mutations per gene per individual). As mutation in microsatellites is due to slippage during replication (around 10^{-3} mutations per gene per individual) rather than point mutation, it would not be safe to use Nei's formula (time of divergence = 5×10^6 *D*) to extrapolate any times of divergence between taxa based on microsatellite data.

6

Fish Farming and Aquaculture

Introduction

Fish farming is the principal form of aquaculture, while other methods may fall under mariculture. Fish farming involves raising fish commercially in tanks or enclosures, usually for food. A facility that releases juvenile fish into the wild for recreational fishing or to supplement a species' natural numbers is generally referred to as a fish hatchery.

Worldwide, the most important fish species used in fish farming are carp, salmon, tilapia and catfish. There is an increasing demand for fish and fish protein, which has resulted in widespread overfishing in wild fisheries.

Fish farming offers fish marketers another source. However, farming carnivorous fish, such as salmon, does not always reduce pressure on wild fisheries, since carnivorous farmed fish are usually fed fishmeal and fish oil extracted from wild forage fish. In this way, the salmon can consume in weight more wild fish than they weigh themselves. The global returns for fish farming recorded by the FAO in 2008 totalled 33.8 million tonnes worth about $US 60 billion.

Fish Farm Operations

Once the farmer has found out the place for the pond the next step is to decide what kind of fish culture is possible with the space available. He must also decide whether his resources will allow him for starting a fish farm. This planning is necessary to determine the number of fish ponds to be constructed and the kind of fish culture one wants to take up.

A range of ideas are presented in this section of the booklet concerning the kinds of fish farm operations whether raising fish or breeding fish, whether raising carps, tilapia or catfish, whether prawns or shrimps, etc. He also gets an idea on the type of ponds small or big, grow in one pond or in more ponds, advantages of small and big ponds, advantages of mixing or separating different breeds of fish and sexes, etc.

A farmer who prepares himself to start fish farming should keep in mind that a certain percentage of fish introduced into the pond will die however much we try to take care of them.

In the nature many fishes never reach their adult size because they are eaten by other animals or big fishes or they die of diseases. "Survival of the fittest" is the law of nature. In the artificial rearing of fishes in the ponds the farmer tries to provide maximum growing and development conditions for getting maximum number of fishes and thereby maximum number of fishes reach marketable size.

Kinds of operations

As already mentioned once the site is identified the farmer has to decide whether he wants to go for breeding and production of fry and fingerlings or rear them till they attain marketable size. For breeding he needs more number of small ponds. The number of ponds may be more if he is trying to breed several breeds of fishes and in the same breed several sets of breeders (one set of breeders consists of two males and one female). If the farmer is planning to rear the fish only he needs big ponds. Depending on the number of fishes he wants to rear he may need one or more ponds. He also should consider whether he is raising the fish for the family usage or tor marketing purposes and whether he wants it as a main business or as a subsidiary business.

Questions like these have to be considered so that he can construct

- The right kinds of ponds,
- The right number of ponds and
- Stock the right kinds of fishes.

Types of ponds

The types of pond a farmer can build depend upon water supply, soil and topography. Generally, two types of ponds are built -barrage ponds and diversion ponds. The main difference between these two types of pond is the water source though in many aspects the construction details are the same.

Barrage Ponds

In barrage ponds the water source is the rain run-off water or a spring forming into a small stream running down the valley. The site is marked by a valley type of topography and the pond is constructed by building a wall across the valley and store the running water through the valley. The area where water is collected is called the pond area.

The barrage pond requires only one wall at the lower side of the pond area. By building a number of walls at suitable intervals the farmer can make several ponds successively one after the other forming a chain of ponds. The water enters first into the pond which is at the highest elevation among all the ponds and then passes on to the next pond lower in elevation. In other words the drained out water from the higher pond is collected by the lower one through a drainage pipe and then to the next lower till we reach the last one. From the last one the water is drained out safely. When ponds and the water inlet and outlet are arranged in such a way that the water flows from one pond to the other it is called ponds in series or rosary system.

Water can also be directed independently into each pond from a channel running along one side of the pond and drained out into another channel at the other side of the pond. This is the best system of water supply to the fish ponds. Arrangement of ponds in this way is called ponds in parallel system. The channel that supplies water is called inlet channel and the channel that collects water from each pond is called drainage channel.

The drainage outlet may be fixed at the bottom side of the pond if the pond is to be drained out completely. Besides this drainage outlet there should bean outlet at a height beyond which the farmer does not want to accumulate the water in the pond. This is for the protection of the build which will be broken if water gets accumulated beyond certain height. The excess water is drained out from the pond through this drainage which acts as an overflow channel.

The drainage and the overflow channel should be constructed in such a way that the water should be drained out safely without eroding the soil and causing damage to the pond. In barrage ponds, flow channels are fixed at a suitable height in order to drain out the excess water that may be accumulated during the rain. There are many variations of overflow channel system from which the farmer can choose the most suitable one. The drainage and overflow channel can be joined together just before they join the common big drainage channel.

Barrage ponds should not be built where the flow of the water is high. The walls will not be able to withstand the force of rushing water especially during the rainy season.

Diversion Ponds

Diversion ponds are those which receive water from river or stream through a diversion channel. One diversion channel takes water to the pond and another diversion channel act as a drainage. The drainage channel may be taking the water into the same river or stream from which the water came in or to a farm where it is used for irrigation or the water may be drained out into any other outlet. If more than one pond are constructed they are connected with the inlets and outlets in series or in parallel design. Diversion pond can be made in a number of ways depending on the ingenuity of the farmer. It can be built anywhere provided water can be taken to it from a water source by channel or pipe. If the pond is built on a flat ground there should be walls all around to prevent the collapse of the sides. Sometimes the ponds are constructed by deepening an area which is lower than the surroundings. If there is enough space and water supply is assured there can be several ponds successively constructed to use the overflow of water from one diversion pond to a second one and from the second to the third and so on, till the last.

Whatever may be the type of pond being constructed there will always be some advantages and disadvantages. The farmer has to consider both the advantages and disadvantages and make his decision most beneficial to him. However, diversion ponds are better than barrage ponds because they are less likely to overflow and the water source is more dependable. But barrage ponds are cheaper in construction. Similarly parallel system of pond construction is better than the se- ries or rosary system. One can have a mixture of barrage ponds and diversion ponds arranged in series (rosary) or parallel system.

The art of planning and constructing fish ponds is very much an individual thing. There are many ways of using land and water resources. But the exact shape, number and size of the ponds have to be decided by the farmer.

Number of Ponds

The number of ponds usually depends on the number of functions the farmer wants to perform and also the number of times he would like to repeat a single function within a given period of time. The functions related to the fisheries are holding a breeding stock, breeding, hatching, raising fry, raising fingerlings and rearing for marketing. Normally these functions are better done in different ponds or water bodies.

At the same time it should be remembered that several functions can be performed in the same pond or water body. For example, breeding and hatching and raising fry can be done in the same pond. However, for breeding large number of breeding stock we require more than one pond. Similarly, the number of ponds also depends on the number of breeds or kinds of fish the farmer wants to breed. For example, if the farmer wants to breed several breeds of fish, different ponds may be required, or several sets of the same breed is to be maintained. In the same way if the frequency of breeding is more (eg. once a week, once in two weeks, a month, etc.) the number of ponds required will be more. At least two ponds are required for rearing for market purposes. If the farmer cannot make more than one pond he can also breed and hatch the fish in an enclosure of inverted mosquito-net fixed inside the pond.

Size of the Pond

The size of the pond also depends on topography, water supply, and the need or volume of production one intends to have. However, the nursery ponds will be smaller than the rearing ponds. Apart from this the size of the rearing pond itself can be varying. The advantages of small ponds are, function like harvesting, draining, refilling, treatment for diseases, etc., can be done quickly and easily. But the cost of construction and the land area required for the small ponds will be more compared to large ponds. But for the small and marginal farmers few small ponds are better than having large ponds.

Depth of the Pond

The depth of the pond also depends on several factors such as breeding, fry raising, fingerling raising, rearing for marketing, etc. Least depth is required for breeding and fry raising and maximum depth is required for rearing for marketing. Normally for breeding and fry raising the pond depth ranges from 0.5 to 1.0 m while for raising fingerlings a depth of 1.0 to 1.5 m is recommended. For rearing fish to market size the recommended pond depth varies from 1.2 to 2.0 metre. Normally, sunlight does not reach deeper than two metres and the growth of the planktons will be restricted. At the same time too shallow pond will heat up the water or too much vegetative growth may occur and consequently the available oxygen to the fish will be much less than the required especially during the night hours. But in commercial farms breeding and raising of fry is done in plastic tubs, basins, glass jars or even in big mud pots. Glass vessels are better for this purpose because we can observe the fertilization, hatching and the conditions of the fry well.

Single Pond Operation

If the farmer can have only one pond he could purchase or collect fry and fingerling and rear them to market size. For this there should be regular supply of fry and fingerlings. He can also breed fish inside a netted area in the same pond.

The farmer who is having single pond can go for culture of single breed or a mixture of several breeds. The former is called monoculture and the later polyculture. Both have advantages and disadvantages.

Monoculture

Monoculture is the fish culture of only one species. It is intensive in caring, feeding and other management practices. It is easier to perform all the management practices since only one type of fish is involved. In monoculture the farmer has greater control over the size, age and sex.

In mono culture fish can be harvested selectively by using nets of different mesh size. In monoculture diseases and pests are easily controlled at the same time there is a possibility of complete elimination of the whole lot by some epidemic type of diseases and pests if they go out of control.

Polyculture

In polyculture different breeds of fish are mixed together in fixed proportion or at random. Polyculture is more closer to the natural method of growing fish and under this situation only the fittest ones will survive. Several small fishes may be eaten by the bigger ones or they could not compete with the bigger ones for food and space. This problem is more in the case of poly culture in which both carnivorous and herbivorous fishes are reared together.

At the same time a farmer can have a polyculture with proper combinations (in number) of selected breeds of fishes. The selection is done on the basis of feeding habits. Such polyculture is called composite fish farming. For example, a farmer can rear a mixture of Indian carps such as catla, rohu, mrigal and Chinese carps such as silver carp, common carp and grass carp.

The catla and silver carp normally takes feed from the upper layers of the water body, the rohu and common carp from the middle layers, while the mrigal and grass carp takes feed from the bottom layers of the pond. All are herbivorous in their feeding habit. In the composite fish culture full utilisation of the pond is possible.

If the breed introduced are naturally fast multiplying like tilapia and their number is a problem in the pond, another fish breed like catfish which is carnivorous can be introduced. The catfish will feed on the organisms at the bottom as well as the fry of tilapia keeping the population of tilapia under control.

Introducing some grass carp among other breeds of fish that are being reared is the best method of controlling the weed growth in the pond. For farmers who have only one pond and do not have facilities like regular supply of fry and fingerlings nor he is unable or unwilling to breed them and those who cannot feed and care in the scientific and intensive ways, it is advisable to go for polyculture with fishes that naturally breed in the pond.

Monosex culture

In monosex culture either male or female fish of the same breed or more than one breed is reared. The advantage is faster growth of the fish since all the energy is concentrated on the growth and development. No energy is used for the reproduction. Sexes can be separated individually during the breeding season or by noting the difference between the male and female in a breed of fish. In the cross breeding of some fish the off springs produced are sterile by nature or produce hundred per cent male fish. For example, if we cross male tilapia of *Tilapia macrochir* with female of *T. nilotica*, the male of *T. mossambica* with the female of *T. nilotica* and male of *T. hororum* with female of *T. mossambica* the off springs will be completely male. There are no crosses in which 100 per cent off-springs produced are female. Monosex culture is valuable but difficult for ordinary farmers to carry out the breeding and rearing. Therefore, it is not recommended generally to the small scale fish farms unless there is a steady supply of male fry or fingerlings.

More than one pond operation

If the farmer has more than one pond then, he can rear fry or fingerlings for marketing in all the ponds or he can take up the breeding in one, raise fry or fingerlings in another and can rear fish up to marketable size in another pond. If he still has more ponds he can replicate the same functions in more than one pond. With three ponds a farmer can go for breeding and raising of fry for his own farm. The smallest one is used for breeding while the others may be used for rearing till they attain marketable size.

The major difference between a large farm operation and a small farm operation is the number of ponds and the fry or fingerlings reared. In big fish farms all the input requirements will be more so also the greater output can be expected in terms of production as well as income.

Types of Integrated Fish Farming

Basically the integrated fish farming is of two types:

1. Agri-based fish farming
2. Live-stock fish farming

The fish-cum-live-stock farming is realised as innovation for recycling of organic wastes as well as production of high class protein at low cost.

Agri-based fish farming

Paddy-Cum-Fish Culture

In India, this farming is practised in the states of Bihar, West Bengal, Orissa and Assam where enough water is present in the paddy fields. The paddy fields retain water for 3-8 months in a year. The interest in this practice has declined in recent years due to the use of pesticides to protect high yielding varieties of paddy.

This practice can be done in following types of paddy plots-

- Perimeter type–paddy grows in the middle.

- Central pond type – paddy growing area is on the perimeter.
- Lateral trench system–trenches are provided on either one or both sides of the moderately sloping field.

The variety of rice used in this culture is Panidhan, Jalmagna, CR26077, Tulsi, etc., while the fish spp. are Indian major carps, Channa spp, Oreochromis mossambicus, Clarias batrachus, Anabas testudineus, silver carp, grass carp, common carp. The total production in such practice is approximately 90 quintal from 2 paddy crops while the fish production is about 1000 kg from 1 ha.

Horticulture-Cum-Fish Farming

The horticulture-cum-fish farming system includes the culture of fruits, vegetables and flowers on the embankment of the pond. The fruits and vegetables contain various nutritive elements and the Indian Council of Medical Research has recommended 85g of fruits and 300g of vegetables to consume daily. For horticulture crop production the inner and outer dykes of the pond and adjoining areas are used. The selection of plant is the main criteria for the success of this system. The plant should be dwarf, seasonal, evergreen, remunerative and less shady. The fruit crops which can be used are Mango, Banana, Papaya, Coconut, Lime, etc., and the vegetables like Brinjal, Tomato,Cucumber,Gourds, Chilli,Carrot, Raddish, Turnip, Spinach, Peas, Cabbage, Cauliflower, Ladies finger can be grown according to their season throughout the year. The flower plantation on the embankment is also useful. We can use the plants like Rose, Jasmine, Gladiolus, Marigold and Chrysanthemum, etc., which provides additional income to the farmer and beauty to the farm. This system provides 20-25 per cent more return in comparison to aquaculture alone.

The agri-based fish farming includes the mushroom fish system, sericulture-fish system, fodder crop integration, etc. Pond bundhs may also used for growing pulses and oil seed crops. Aquatic cash crop like Makhana (Euryale ferox) and Singhara (Trapa spp) integration can also be done along with air-breathing or carnivorous fishes.

Live-stock fish farming

Poultry-Cum-Fish Farming

This system utilises poultry droppings of fully built- up poultry litter for fish culture. The fish production obtained is about 5000 kg/ha/yr. with 1250 kg chicken meat and 70000 no. of eggs. Approximately 500-600 no. of birds is reared in a 1 ha pond. The Rhode Island or Leghorn variety birds are more preferred over others. They require 0.3-0.4 square meterspace/ bird. Hoppers are used to feed them and to minimise feed wastage. The poultry birds (layers) are fed with starter, grower, and brooder feed according to their age. In India it is practised in Andhra Pradesh, Bihar, Haryana, Kerala, West Bengal, Uttar Pradesh, Maharashtra, Orissa, and Tamilnadu.

Duck-Cum-Fish Culture

The duck are commonly called as biologicals aerator. They are reared on the dyke of the pond in a low-cost house. This farming is practised in Tamilnadu, Assam, Bihar, Andhra Pradesh, Tripura, Orissa, Karnataka, Kerala and Uttar Pradesh. The 'Indian runner' and 'Khaki campbell' varieties are found more suitable in this culture. About 300 no. of ducklings (some spp. are reared 450-500 in no.) are reared to fertilize the 1 ha. pond. The duck not only act as live aerator by splashing water with their webbed feet but also control the aquatic weed (Lemna, Azolla, etc.), aquatic insects, molluscs, tadpoles, etc. Each duckling require about 0.3-0.5 square meter area as living space. The total production from such type of culture is about 3500-5000 kg fish, 18000-18500 eggs and 600 kg of duck meat. The duck droppings are used as manure for primary production.

Pig-Cum-Fish Culture

This system has certain advantages over others. The 30-35 pig's waste may produce 1 tonne of Ammonium Sulphate and 40-45 pigs are adequate to fertilize 1 ha water area under polyculture. Each pig requires about 3-4 sq.m floor space. This system of integration is very common in China, Taiwan, Vietnam, Thailand, Malaysia, Hungary and some European countries.

The White Yorkshire, Hampshire and Landrace are the popular breed of pig for integration with fish. Pigs need clean housing which should provide adequate protection from adverse climates. The pigs are fed on pig mash which is made up of rice bran, rice polish, wheat-bran, broken maize, ground-nut oil cake, fish- meal mineral mixture, salt, etc. The spoiled vegetables can also be mixed in it. This system provides about 3000-4000 kg/ha/yr fish, 4500 kg/yr pig meat and 800 no. of piglets every year.

Cattle-Cum-Fish Culture

It is a common practice all-over the world. The cow excreta is most abundant in terms of availability and a healthy cow may excrete over 4000-5000kg dung and 3500-4000 litre urine on an annual basis. The BOD of cow manure is lower than other livestock manure. About 5-6 cows can provide adequate manure for 1 ha pond in addition to 9000 kg milk and about 3000-4000 kg of fish annually. Cow-shed should be built close to fish pond to simplify handling of cow-manure.

Goat-Cum-Fish Integration

It is considered as poor man's cow and a goat's excreta is considered as a very good organic fertilizer. The goat excreta contains organic carbon-60 per cent, N-2.7 per cent, P-1.78 per cent, K-2.88 per cent and its urine is also equally rich in both N & P. At least 50-60 goats are essential to fertilize 1 ha pond. The goats should be provided with dry, safe, comfortable house protected from excessive heat. The goat breeds are Jamanapari, Beetal, Barbari for milk and Bengal, Sirihi, Deccani are used for meat purpose. Goats are selective feeders and consume Berseem, Napier grass, Cowpea Soybean, Mulberry, etc. This integration can provide 3500-4000 kg fish/ha/year without supplementary feeding and fertilizer.

Rabbit-Fish Integration

Rabbit meat is preferred by most of the health conscious consumers owing to its low fat in comparison to other meats. The important meat breeds are Soviet Chinchilla, Grey Giant, and White Giant, etc. Rabbits are reared in cage, hutch and floor system (floor should be cemented). Rabbit excreta contain organic carbon-50 per cent, N-2 per cent, P-1.33 per cent, and K-1.2 per cent. The rabbit excreta is high in nitrogen content and low in moisture, thus quality manure for sustained plankton production. It is estimated that excreta from 300 rabbits would be enough for 1 ha pond fertilization.

Major categories of fish aquaculture

There are two kinds of aquaculture: extensive aquaculture based on local photosynthetical production and intensive aquaculture, in which the fish are fed with external food supply.

Extensive aquaculture

Limiting for growth here is the available food supply by natural sources, commonly zooplankton feeding on pelagic algae or benthic animals, such as crustaceans and mollusks. Tilapia species filter feed directly on phytoplankton, which makes higher production possible.

The photosynthetic production can be increased by fertilizing the pond water with artificial fertilizer mixtures, such as potash, phosphorus, nitrogen and micro-elements. Because most fish are carnivorous, they occupy a higher place in the trophic chain and therefore only a tiny fraction of primary photosynthetic production (typically 1 per cent) will be converted into harvest-able fish.

Another issue is the risk of algal blooms. When temperatures, nutrient supply and available sunlight are optimal for algal growth, algae multiply their biomass at an exponential rate, eventually leading to an exhaustion of available nutrients and a subsequent die-off. The decaying algal biomass will deplete the oxygen in the pond water because it blocks out the sun and pollutes it with organic and inorganic solutes (such as ammonium ions), which can (and frequently do) lead to massive loss of fish.

An alternate option is to use a wetland system such as that of Veta La Palma.

In order to tap all available food sources in the pond, the aquaculturist will choose fish species which occupy different places in the pond ecosystem, *e.g.*, a filter algae feeder such as tilapia, a benthic feeder such as carp or catfish and a zooplankton feeder (various carps) or submerged weeds feeder such as grass carp.

Despite these limitations significant fish farming industries use these methods. In the Czech Republic thousands of natural and semi-natural ponds are harvested each year for trout and carp. The large ponds around Trebon were built from around 1650 and are still in use.

Intensive aquaculture

In these kinds of systems fish production per unit of surface can be increased at will, as long as sufficient oxygen, fresh water and food are provided. Because of the requirement of sufficient fresh water, a massive water purification system must be integrated in the fish farm. A clever way to achieve this is the combination of horticulture and water treatment, see below.

The exception to this rule are cages which are placed in a river or sea, which supplements the fish crop with sufficient oxygenated water. Some environmentalists object to this practice. Expressing eggs from a female rainbow trout per unit of fish weight Selenium is higher than in extensive farming, especially because of the high cost of fish feed, which must contain a much higher level of protein (up to 60 per cent) than cattle food and a balanced amino acid composition as well. However, these higher protein level requirements are a consequence of the higher food conversion efficiency (FCR-kg of feed per kg of animal produced) of aquatic animals. Fish like almon have FCR's in the range of 1.1 kg of feed per kg of salmon whereas chickens are in the 2.5 kg of feed per kg of chicken range.

Fish don't have to stand up or keep warm and this eliminates a lot of carbohydrates and fats in the diet, required to provide this energy. This frequently is offset by the lower land costs and the higher productions which can be obtained due to the high level of input control.

Essential here is aeration of the water, as fish need a sufficient oxygen level for growth. This is achieved by bubbling, cascade flow or aqueous oxygen. Catfish, Clarias spp. can breathe atmospheric air and can tolerate much higher levels of pollutants than trout or salmon, which makes aeration and water purification less necessary and makes Clarias species especially suited for intensive fish production. In some Clarias farms about 10 per cent of the water volume can consist of fish biomass.

The risk of infections by parasites like fish lice, fungi (Saprolegnia spp.), intestinal worms (such as nematodes or trematodes), bacteria (*e.g.*, Yersinia spp., Pseudomonas spp.), and protozoa (such as Dinoflagellates) is similar to animal husbandry, especially at high population densities. However, animal husbandry is a larger and more technologically mature area of human agriculture and better solutions to pathogen problem exist. Intensive aquaculture does have to provide adequate water quality (oxygen, ammonia, nitrite, etc.) levels to minimize stress, which makes the pathogen problem more difficult.

This means, intensive aquaculture requires tight monitoring and a high level of expertise of the fish farmer. Very high intensity recycle aquaculture systems (RAS), where there is control over all the production parameters, are being used for high value species. By recycling the water, very little water is used per unit of production. However, the process does have high capital and operating costs. The higher cost structures mean that RAS is only economical for high value products like broodstock for egg production, fingerlings for net pen aquaculture operations, sturgeon production, research animals and some special niche markets like live fish. Raising ornamental cold water fish (goldfish or koi), although theoretically much more profitable due to the higher income per weight of fish produced, has never been successfully carried out until very recently. The increased incidences of dangerous viral diseases of koi Carp, together with the high value of the fish has led to initiatives in closed system koi breeding and growing in a number of countries. Today there are a few commercially successful intensive koi growing facilities in the UK, Germany and Israel.

Some producers have adapted their intensive systems in an effort to provide consumers with fish that do not carry dormant forms of viruses and diseases.

Specific types of fish farms

Within intensive and extensive aquaculture methods, there are numerous specific types of fish farms; each has benefits and applications unique to its design.

Fish Rearing in Cages and Pens

As fish necessarily requires water, most people tend to think that for fish rearing they should have their own fish pond which may be situated near the streams, rivers, lakes, permanently flooded areas or reservoirs but do not have the ownership of any portion of any of them, or he may be having a water body which is linked with a lake, reservoir or river. Under such conditions one can go for rearing of fishes In cages and pens. The idea is to rear fish in a common and unbounded water body. This booklet explains the basic methods employed on cage and pen fish farming. Many poor farmers may get benefit from this method, at least partially meet their food and finance requirement.

If a farmer is living near a huge water body, such as canal, river, stream, lake, permanently flooded area or reserviors which does not belong to him or he has a water body which is linked with some other huge water bodies (rivers with lakes or sea, streams with big rivers or flooded areas) he can rear fishes in cages and pens. Of course under this system only a limited number of fishes or a specific type of fish may be reared. However, the number will depend on the size of the cage or the pen used. Those who are able to establish bigger pens and cages will be able to rear proportionately large number of fish provided he can feed them.

Feeding of fish reared in cages and pens is very important among the management practices. Therefore the main thing one should consider is the sufficient availability of feed for the fish reared in the cages and pens. Of course the type of feed depends on the type of the fish being reared Some fish are carnivores, some herbivores and some others are omnivores. For more details about feeding of fish please consult Booklet No. 525 on "Fish Feeding". Generally, a mixture of oil cakes and rice polish is a very good feed for most of the species of the fishes that are being reared by people.

The cages and pens can be constructed with any material conveniently available at the local level provided it is durable at least for one year. However, more the durability of the material greater will be economic benefit from this system of fish rearing though the initial cost may be more.

This type of fish rearing has been practiced since long in certain parts of the Asia. Perhaps this system of fish culture is originated with the view to control retrieval of the stocked fish. Its application is more relevant for those people for whom the suitable water supply and land for fish culture are becoming less and less available. Further to intensify this system of fish culture, scientific methods are being evolved for increasing the production of fish in cages by forced aeration or by using different types of filters and better feeding and management practices. In India this system can be practiced in the low lying and permanently flooded areas of Kerala, West Bengal, Orissa, Andhra Pradesh, Karnataka, Goa, Gujarat, North Bihar, Assam, Manipur, etc., though much work needs to be done still to perfect and standardise this system.

Structure of Cages and Pens

The cages are structures which are covered and secured from every sides and the fishes are kept inside. They can be fixed stationary, suspended, partially or fully immersed, or floating, depending on the convenience. The cages are fitted with a door which can be opened in order to introduce the fishlings, to feed and care them and to harvest them at the end of the rearing season.

The best positioning of the cages is to fix them in such a way that the top portion of the cage is exposed to air and sunlight so that the fish can breath air whenever it is necessary. Also due to the constant contact with the atmospheric air the amount at oxygen dissolved into the water will be more to the benefit of the fishes.

On the other hand, the pens are fixed as fencing structures or enclosures covering an area of the water body. They are made by fixing the fencing material into the ground at the bottom and rising right through the water column and make them stand at about 1-2 ft above the water. The pens are open at the top and exposed to atmosphere. They cover much larger area than the cages. In a way pens are enclosed areas.

Pens also give the idea of several sections of an enclosed area each harbouring a certain number of fishes of different type, size, and age. For breeding purpose the breeding stock may be kept in small pens. Similarly for spawning and hatching also separate pens are required.

The pen looks like a fenced area. Wooden reapers, bamboo splits, reeds or any other suitable material can be used for the construction of the fence. The fencing material is fixed vertically on the ground, under the water and held very close to each other by ropes. Horizontally fixed fencing material is so closely arranged that the gap between them is reduced to the minimum required space, that would not allow the fish to come out and also they will remain safe from predators.

Enclosed area denotes a larger area than the pen area. A portion of large water bodies such as lakes, reserviors, flooded areas, etc., is fenced around in order to rear large number of fish for commercial purpose as is practiced by some people living near the Chilka lake in Orissa.

The pens and enclosures are established only in shallow waters whereas the cages can be used even in deep water areas.

Material for Construction

There are a number of materials that can be used for making the cages and pens for fish rearing. Depending on the ingenuity of the farmers and availability of the things a farmer can use any type of material to construct the cages and the pens. Some of the common ones are mentioned here.

Bamboo and Coir Rope

Mature and seasoned bamboo splits tied closely using firmly twisted coconut fibre rope perhaps is the best material for making cages and pens in the rural areas where other commercially produced materials are not usually available. Both these materials are durable in submersed conditions. Thin bamboo splits of half inch thickness and width are held together closely by inter-twining them using thin coconut fibre ropes. The main frame of the cage is made of bigger bamboo splits or wooden reapers.

Nylon Threads

Strong nylon threads are better than any other material for making the cages and pens though this may be quite costly for an ordinary farmer. Nets are made out of this nylon thread. The mesh size of the net depends on the size of the fishlings which will be reared in the cage. For fry (up to 15 days old) the mesh size should be very small, for the fingerlings (15 to days) the size will be bigger, while for the growing or stocking fishes (reared for market size) the mesh size will be still bigger (half to one inch size).

Plastics

Tubes, sheets, ready made meshes made of strong plastic material can be used as cage and pen material. However the plastic should be very strong and non- degradable. Now-a-days very strong and non- degradable type of plastics are available in the market. One has to search and find out the most suitable material for his own purpose.

Wire Meshes

Wire mesh is another important material for the construction of cages and pens for fish farming. The mesh as already mentioned could be of any size depending on whether the farmer wants to keep the fry or fingerlings or fish grown for market purpose. It is highly advisable that the mesh is made of galvanised wire in order to reduce the possibility of rusting. Rust can be also controlled by coating with water proof paints on the wire mesh.

Wood

Wood can be used as a caging. or penning material. There are several types of wood which can withstand the submersible condition. There could be many other materials which are locally available and could be used for the construction of the cages and pens for fish rearing. The size and shape of the cage or pen is decided by the farmers. Those who have large space and want to rear fish in large number will need bigger cages and pens. But they could be made in any shape and size. Depending on the space available the cages and pens may be constructed big, medium or small, long or short. The shape may be rectangular, square, triangular, irregular or any other: The size of the cage and pen too vary a lot tram very big which harbours hundreds or thousands of fish, to very small in which only one or two fish may be kept.

Others

There may be several other materials which may be available locally or commercially for the construction of the cages and pens. It is for each one to look for the appropriate material available at cheaper rate and to use them for the construction.

There may be ready made modular type of material in the form of parts of the cage or pens. Using them we can construct, dismantle and reconstruct the cages and pens. Such materials may not be possible at the farmer's level in the case of cage construction but it is possible in the case of pen construction.

Construction and Fixing of Cages and Pens

Different methods are followed for the construction of the cages and pens for fish rearing. They are explained as follows under separate sub-heads. The construction of enclosures is very much similar to the construction of the pens. Only the enclosures are constructed in a bigger area while the pens cover comparatively a smaller area.

Construction and Fixing of the Cages

First of all the site where the cage is to be fixed is surveyed and measured to determine the size, shape and the way the cage is going to be fixed at the location. Arrangements for the materials are done accordingly.

The cage may be constructed in situ or constructed in parts at a different place.and brought them to be assembled at the place where It is going to be fixed. The most important point to be kept in mind is the mesh size. The mesh size is made bigger for keeping the bigger fishes whereas for the small fish smaller mesh work is provided.

The cages are fixed in immovable or in floating manner. They can be fixed completely or partially immersed in the water. The cages are fitted with sufficient doors which will help in the feeding, caring and harvesting of the fishes grown. The cages are fixed in such a way that always the fresh water will flow through them and also enough sun light is available so that the fish inside will get the most optimum conditions to live.

Construction and Fixing of Pens

As in the case, first of all of cages the location is surveyed and the type and the amount of material needed is estimated before the construction of the pens and enclosures. The construction of the pens and enclosures are generally done in situ in the form of a fence. The fencing material consists of the main pole fixed firmly in the ground under the water at convenient distances. These are jointed by at least two horizontal bars or reapers one on the upper side and other on the lower side. If needed a third bar or reaper is fixed horizontally between the above mentioned horizontal bars. On to these horizontal reapers or bars are fixed the vertical bars closely to form the fencing for the pen or enclosure.

The vertical bars are fixed on the horizontal bars using nail or by tying with strong rope material. The distance between the two vertical bars will be adjusted to hold the minimum size of the fish that are released into the pen or enclosure. The pens can be constructed by fixing different segments of a fence. Each segment consists of a fixed number of vertical bars closely fitted by using two or three horizontal bars. The vertical bars at both ends and at the middle will be longer by one foot and pointed Because of these three vertical bars a segment can be fixed vertically by driving them into the soil till the other bars in the segment come into close contact with the soil surface.

After fixing the first segment fix the second segment close to the first one; then fix the third, fourth segments and go on till the whole fencing is completed to create an enclosed area. Using the same technique we can also separate the whole enclosure into several sections if needed. But even in fixing the segmented fences also we need the strongly fixed poles on to which the segments are fixed firmly. Pole can be fixed between each segment or at least between two segments each.

We can have segments without the above mentioned three longer and pointed vertical bars on each segment. In this case the segments are pressed down into the soil between two poles already fixed at an interval equal to the width of the segment and tied to tern. The pens are fitted with a door made of the same material as the pens. The door will be more like a segment of the fence which can be removed by lifting when one wants to open and fixed back when one wants to close it. The door is better secured on to the fence by tying rather than fixing it with hinges. Movement of the door fixed with the hinges does not work under the water and the door has to remain fixed into the soil. However one can adopt several ways to fix a door and it should be fixed in such a way that the door can be opened and closed like any other door fixed on hinges. The depth of the cages or the depth of the water in the pen should be sufficient for the fish to grow. The optimum depth for cage varies from 60 to 120 cm, and for the pens the depth of the water inside the pen should be at least 120 cm up to a maximum of 180 centimetres. The length of the cage should be minimum ten feet and the width should be at least three feet. However depending on the ability of the farmer the cages may be bigger or smaller than the above mentioned dimension. But bigger the cages better they are for the growth of the fishes. Fish require space for swimming around and the more we give them space greater will be the growth rate.

In the case of pens the only recommendation is to provide as much area (length and width) as possible. Obviously the depth, will be the same as that of the water body in which the pen is constructed. For the purpose of, hatching and fry raising cages and pens are unsuitable as the mesh size is bigger and the try can escape from the cages and pens. For this the cages and pens should be fitted with fine mesh net as that of a mosquito net. One can also use "Happas" (enclosures made of fine mesh net like the inverted mosquito net) for breeding, hatching and rearing of try up to the age of 90 days, by which time they grow up to the size of a fingerling. If the mesh size of the cage or the pen is sufficiently smaller, the fingerlings can be released into the cages or pens. Otherwise they have to be kept in suitable enclosures till they are grown big enough to be contained inside the cage or pen. Details about breeding and hatching of the fry and raising up to fingerlings stage are described in Booklet No. 543 on "Carp Breeding".

Types of Fishes Reared in the Cages and Pens

There are a number of fishes that can be reared in cages and pens. In fact any type of fishes, prawns, frogs, crabs and shell fishes can be reared in the cages and pens. Some of the fishes that are suitable for cage rearing under mono-culture are: the air breathing fishes such as murrel (Channa marulius), striped murrel (Channa striatus), spotted murrel (Channa punctatus), magur of catfish family (Clarias batrachus), singhi (Heteropneustes tassilis), the fishes of tilapia family, the fishes of Indian carp family (catla, rohu and mrigal) and the Chinese carp family (silver carp, grass carp and common carp), trouts, etc.

Depending on the availability and the location any type of fish can be reared in the cages or pens. Since the cage arid pen rearing involves considerable amount of time, energy and finance it is advisable to rear the fast growing type of fishes rather than local types which are slow growing. Among the types of the fish above mentioned, grass carp needs special mention as they can be reared easily on vegetation. In places where there is plenty of marshy and water logged lands there air breathing fishes are most suitable for cage rearing In poly culture a combination of tilapia and catfish is more suitable for a poorly managed pen or enclosure conditions. They are also suitable to the cage conditions. Tilapia is a fast multiplying type of fish which is controlled by the carnivorous catfish.

In a well managed cage any single type of the carp fishes can be reared. The number of fish reared depends on the volume of the cage. A cage is too small for composite fish, farming (rearing a number of fishes together), but in the pens and enclosures one can go for composite fish farming as they replicate in the pond situations to a great extent. Rearing of frogs, prawns, shell fishes, etc., is not yet a common practice in India and a lot more research work needs to be done. The readers are aware that all types of ornamental fishes are reared in the cages for which a separate booklet is being prepared.

Management Practices

After the cages and the pens are properly constructed the fishes are introduced into the cages or pens. The fingerlings may be raised in one's own Lann or may be bought from some reliable nursery. Only the fingerlings bigger enough to be contained inside the cages or pens, are introduced. Others may be kept in special enclosures till they grow big enough to be introduced into the cage or pen. In the cage and pen system one can adopt monoculture or polyculture depending on one's liking, availability of the feed, availability of the fry, etc. The number of fish reared per cubic metre volume depends on the size of the fish. Roughly we can say that one kilogram size fish per cubic metre volume of water may be held as a thumb rule. This is the minimum space needed. Accordingly two fish of half a kilogram each or four fish of 250 gm each or eight fish of 125 gm each or twenty fish of 50 gm each or forty fish of 25 gm each can be grown in one cubic meter volume of water. Though such mathematically arranged numbers may not always be feasible still such considerations gives one an idea of the number of fish that can reared in a given volume of cage and the pen. In the cage rearing system the fishes do not get any natural feed though in the pen system, depending on the site there may be natural planktons of flora and fauna available; Hence the feeding becomes the most important aspect in fish rearing in the cages and pens. Regular feeding with sufficient quantity and quality feed should be carried out in cage rearing. Feeding should also be taken into consideration the nature of the fish reared whether carnivores or herbivores. Normally a mixture of oil cake and rice or wheat bran form the base in preparing feed for the fish. Along with this fish meal, meat meal, prawn meal, various types of fungi, worm meal, etc., should be included into this mixture may be raised in one's own farm or may be bought from some reliable nursery. Only the fingerlings bigger enough to be contained inside the cages or pens, are introduced. Others may be kept in special enclosures till they grow big enough to be introduced into the cage or pen.

In the cage and pen system one can adopt monoculture or polyculture depending on one's liking, availability of the feed, availability of the fry, etc. The number of fish reared per cubic metre volume depends on the size of the fish. Roughly I we can say that one kilogram size fish per cubic metre volume of water may be held as a thumb rule. This is the minimum space needed. Accordingly two fish of half a kilogram each or four fish of 250 gm each or eight fish of 125 gm each or twenty fish of 50 gm each or forty fish of 25 gm each can be grown in one cubic meter volume of water. Though such mathematically arranged numbers may not always be feasible still such considerations gives one an idea of the number of fish that can reared in a given volume of cage and the pen.

In the cage rearing system the fishes do not get any natural feed though in the pen system, depending on the site there may be natural planktons of flora and fauna available; Hence the feeding becomes the most important aspect in fish rearing in the cages and pens. Regular feeding with sufficient quantity and quality feed should be carried out in cage rearing. Feeding should also be taken into consideration the nature of the fish reared whether carnivores or herbivores. Normally a mixture of oil cake and rice or wheat bran form the base in preparing feed for the fish. Along with this fish meal, meat meal, prawn meal, various types of fungi, worm meal, etc., should be included into this mixture.

The quantity of feed depends on the size of the fish bigger the fish greater will be the amount of feed. The more the intake greater will be the growth of the fish. For more details about feeding refer Booklet No.525 on "Fish Feeding".

Feed should be given on a flat vessel or pans placed in the cage or in the pens. This will avoid the loss of feed and it becomes easy to observe the condition of the fish while feeding. Fishes that feed actively indicate that they are in good health and those who do not eat or are sluggish in eating show that they are not in good physical conditions.

Pan feeding also will give an idea whether the feed given is sufficient or not. The second aspect of the rearing of the fish in the cages and pens is to provide suitable environmental conditions to the fish in the cages and pens. Sometimes over crowding in the cages and pens may result in scarcity of feed and less availability of dissolved oxygen in the water. These negative effects should be over come by regular and sufficient feeding of the fish and allowing sufficient flow of water and enough sunlight through the cages or pens.

Harvesting

Harvesting is done in the cages using a hand net. But in the pens and enclosures hand net will work only if the pen is small. If the pen is big we may have to use a small or medium type net depending on the size of the pen or enclosure.

Compared to any other type of rearing methods cage and pen rearing of fish is the easiest system from the point of view of harvesting of the fishes. In the ponds and lakes when netting is done small and big and even the eggs may be caught and the fry and eggs may be destroyed. But in the case of cages one can easily select individual fish while harvesting and others are left free. However in the pens and enclosures harvesting is done by netting or by some other locally existing means.

The harvested fish are weighed and used for domestic purposes or sold out. It is advisable to maintain a record of the number of fish harvested and the yield by weight. Such records will help one to understand at what level of performance one is operating and how much net profit or loss is incurring.

Fisheries and Aquaculture Technology

The technology employed in aquaculture has developed over many centuries but has done so more rapidly during the last half century. Aquaculture systems, and the technology used, vary from very simple systems, used for family ponds in tropical countries, where production is for domestic consumption, to high technology systems, such as intensive closed systems, like those used for rearing stripped bass. Herbivorous and filter feeding fish, reared mostly in simple systems of small freshwater ponds, however, account for about half of global aquaculture production.

Much of the technology used in aquaculture is relatively simple, amounting to small modifications that improve the growth and survival rates of the target species, such as providing additional food, adding seed animals collected elsewhere, managing water exchange to maintain adequate oxygen levels, and protecting the stock from predators. Greater understanding of complex interactions between nutrients, bacteria and cultured organisms, together with advances in hydrodynamics applied to pond and tank design, have enabled the development of closed systems. These have the advantage of isolating the aquaculture systems from natural aquatic systems, thus minimizing the risk of disease or genetic impacts on the external systems.

Developments in engineering, some learnt from offshore oilrig construction, increase the possibilities for offshore aquaculture using robust cages. Sea ranching, the release of young fish into the wild to improve the harvest in capture fisheries, has also made a start but its long term viability is still to be assessed. Major advances are also being made in the technology of the production of aquafeeds, which generally require the combining of a large number of ingredients into very small feed pellets.

Aquaculture Engineering

Aquaculture draws on well-established engineering fields for most of the design and construction needs of its production facilities. Building earth ponds is similar to building roads - a knowledge of the characteristics of soils and the limits of safe design are the basis of good construction. Similarly, the buildings used for hatcheries and other support activities are no different from those common in the housing, agricultural and commercial sectors.

Sometimes ponds are lined with plastics or other impermeable materials, and here the techniques are similar to those for civil structures such as potable water reservoirs or sludge tanks. The design and installation of water control gates, including in unstable soils, benefits from the long experience in this field in the agriculture and irrigation sectors. An understanding of hydrodynamics allows ponds and tanks to be built with good water circulation, oxygen mixing and without 'dead spots' where sediments might accumulate and cause health problems to fish.

For installations in the sea, the situation is somewhat different and many of the important engineering solutions, such as for fish cages or suspended shellfish growout systems, have had to be developed by aquaculturists themselves. They have benefited however from the accumulated knowledge of seafarers in general and fishermen in particular, in the design and operation of mooring and buoyage systems. More recently, when fish farmers have turned their attention to how to operate fish cages in locations further offshore where seas are rougher, the experience of the oil exploration industry has proven very valuable.

Techniques have been developed in recent years for the production of fish and other aquatic products in closed recirculation systems. To make these work, aquaculturists have needed to develop a knowledge of the biological processes operating - such as how bacteria can be used to neutralise and re-cycle the nitrogenous waste products produced by growing fish - and how to engineer the systems to meet the biological requirements. Knowledge of bio-engineering from the waste treatment and water treatment industries has made contributions to the development of closed aquaculture systems and there is probably more that could be usefully transferred from the sewage treatment sector to help solve problems in fish rearing.

Modern materials have brought many benefits to aquaculture. For instance, custom made plastic joints have simplified the construction of sea cages, and made them more reliable in high stress conditions. Experiments with huge free-floating or sunken net cages operated in the open ocean were begun several decades ago, for instance in the Caspian Sea. These showed some promise, but more reliable construction materials will make the farming of fish in such structures increasingly feasible. Modern materials and production methods have been important also, in the construction of plastic filter substrates for indoor recirculating systems. The fine detail of these has been found to make substantial differences to the efficiency of biological filters.In shrimp farms, specially designed matting materials that stand upright on pond bottoms, with a structure that promotes the growth of the small animals and plants that the shrimp can thrive on, have recently been developed and shown to boost production.

Engineering skills are important in the design of most aquaculture facilities and good engineering can affect the efficiency and economics of production. If capital costs can be minimised while still maximising productivity and reducing risk, the farming operation will be more profitable.

Aquaculturists have proven very innovative over the past fifty years, constantly developing new technologies to support their farming operations. As new production methods and species for farming are developed, the engineering solutions needed to support them will continue to evolve.

Aquaculture Facilities

Aquaculture began with man making small modifications to natural habitats so as to improve the survival and growth of target species. Some of the oldest examples are in the rearing of freshwater fish in ponds, which has been practiced for thousands of years in Asia and at least for many centuries in Europe. The simple act of placing a mesh barrier across the outlet of a small pond or lake to prevent fish from escaping can make a big improvement in the supply of food. Similarly, along coasts, some tidal lagoons can be easily turned into ponds. Closing off such naturally occurring water bodies was the start, centuries ago, of much fish and shrimp aquaculture in Asia and, in modern times, in South America. Removing predators and improving the conditions within the pond, (for instance by providing more area of preferred water depth), supplying additional food and, later, by adding seed animals collected outside, were further steps that moved aquaculture production close to where it is today. The farming of seaweeds and molluscs (oysters, clams, mussels etc) developed similarly, as people made improvements such as providing more settlement areas for the young, or the removal of predators from the growing area.

Farmers had a natural desire to improve the productivity of their systems and, as knowledge grew, they learned to stock more animals, increase feeding and manage the exchange of water to maintain the conditions, such as adequate oxygen levels, that the animals needed to survive. The range of facilities used for aquaculture subsequently broadened for a number of reasons. Farmers encountered difficulties with more intensive rearing because of the uncontrolled influence of pond soils, local water quality, weather. Some of these problems could be resolved by rearing in ponds built of concrete, or lined with plastic, by bringing the ponds indoors under cover, or treating the water before flowing it to the culture ponds.

Secondly, as the naturally occurring ponds and lagoons all became used, prospective farmers had to take a broader approach and develop the technology and engineering to be able to use less naturally-favored sites. Net cages floating in protected coastal or inland waters were developed for fish culture, or fish were stocked in fenced areas of the sea or of large lakes. At the same time, the need for supplies of young animals (fry, seed) to stock these systems led to the development of hatchery techniques and dedicated hatchery facilities. For most species this aspect of production has proved more successful when conditions can be more closely controlled, for instance in indoor concrete and fibreglass tank systems, rather than in outdoor ponds. Often, the younger stages of aquatic animals are more sensitive than adults to physical and chemical conditions and these have to be managed within a smaller range, if production is to be successful. Thus hatchery facilities developed as a separate branch of the industry.

More recently, our knowledge has improved greatly regarding the complex interactions that occur in a rearing system, between nutrients, bacteria and the cultured organism. This and technological developments have allowed many aquatic organisms to be reared in completely closed recirculating facilities, including the farming of marine organisms at locations far from the sea. Closed systems have the added advantage of offering greater protection from the danger of disease entering from the natural environment and also of minimizing adverse effects of the production system on that external environment.

Technology has also begun to open up the possibilities of growing fish in enclosures in the open ocean, something that could one day transform the nature of human food production on the planet. With 70% of the earth's surface covered by water, the potential is clear. Earlier technology has restricted the cage farming of fish to sheltered coastal waters. Here the number of available sites is limited, environmental damage is more likely and conflicts exist with other users. As cages are developed that can withstand the demanding conditions of the open ocean, or which can be operated below the ocean surface, farming could move further offshore.

A small start has been made, notably in Japan, through so-called 'ranching' programmes', the farming of the sea without enclosures where fry are released in large numbers into the ocean with the aim of improving returns from the capture fishery. Sometimes artificial reef structures are created underwater as well, to increase the available habitat and natural food for the fish.

Development of Aquaculture Systems

Aquaculture - the growing of aquatic animals and plants - covers a wide range of species and methods. Some of the simplest production systems are the small family ponds in tropical countries where carp are reared for domestic consumption. At the other end of the scale are high technology systems, such as the intensive indoor closed units used in the USA for the rearing of striped bass or the sea cages used in Chile and Europe for growing salmon and bream.

Nearly half of the world's aquaculture however is of herbivorous and filter-feeding fish such as carp. Practised in freshwater, in ponds made of earth, and often managed by a single household, production systems are simple and in many cases have changed little over many centuries. A farmer will excavate a small pond near his house; sometimes, as for instance in the flood river deltas of countries like Vietnam, using the spoil to build a raised foundation for the house itself. The pond will typically be fed by natural rainfall or groundwater, or sometimes by diverting a nearby stream or irrigation canal. Some stocking occurs naturally, but more commonly a farmer will buy young fish from a breeder and stock them in his pond, often after holding them for a time in a small floating net known as a 'hapa' to check that they are in good condition before release in the pond. Production will improve if the farmer can stock a good balance of different species to make the best use of the varied kinds of food available in the pond. Typically the fish will be fed with household or agricultural by-products, and harvested when the family needs a meal. Stocking and harvesting is often continuous, where knowledge of the best stocking level is learned from experience and where the pond is not drained for many years. For most such farmers, fish production is a secondary activity, a useful additional source of protein to add to the supply or income from his main agricultural or commercial activities.

Outdoor freshwater fish farming is practised commercially in many countries, including in the developed nations. The density of stocking may be much higher, the control of feeding, water quality and fish health more closely monitored, but even in intensively operated systems, the key parameters that need to be monitored and balanced for success are similar to those in the traditional household pond system.

Over the last half-century, systems for the indoor rearing of many species of fish have also been developed. Holding of fish in controlled conditions indoors has been important in the development of seed production (hatcheries). Also, as knowledge has increased of nutrient cycles, bacterial action and water chemistry, it has become possible to rear fish, both for food and for ornament, in 'closed' indoor systems, where the water is recirculated. Passing the water through filters that use bacteria to naturally break down and recycle the waste products produced by the fish allows a production system to be run in almost total isolation from the outside. Such systems have been important in the control of disease and also in maintaining the stable conditions that some species need to flourish and reproduce.

A fifth of current world aquaculture production is of plants - mainly seaweeds, that for the most part are grown on the seabed or on raft or racks in shallow coastal waters and used directly for food or for the production of alginate or carageenan (agar-agar). This sector of aquaculture is another largely run by small-scale growers, mainly in Asia but in some parts of the South America, who attach seedlings of marine plants to simple structures built close to shore and constructed from natural materials and then harvest the plants once they have grown.

A further fifth of world production is of molluscs such as oysters, clams and mussels. Again the majority of production is small-scale and by coastal people who often combine fishing activities with farming. Oysters and mussels are mainly grown on structures built above the seabed - poles or racks on the shore, or ropes suspended from rafts or floating lines. The farmer's role is to supply suitable places for seed to settle or to add to natural production by bringing seed from a hatchery - and then to maintain the conditions of waterflow and freedom from predators that the shellfish need to grow. Most commercially grown molluscs feed on microscopic algae floating in the water, so the farmer does not need to provide any feed.

Fish are also grown in coastal areas, in ponds or in floating cages. As technology advances, there is the potential to develop systems for rearing fish in the open ocean, either in sturdy cages or by so called 'ranching', where young fish are released to the wild and then collected by normal fishing or by training them to respond to specially generated sounds.

The tropical coastal zones of Asia and Latin America are where most shrimp farms are found. Shrimp farming is mainly carried out in earth-bottomed ponds built on flat land close to the sea. Shrimp are raised in ponds at a range of different densities. At low densities the systems need only limited inputs of feed and fertilizers. As the density increases, more feed has to be supplied and at the higher end of the range, so-called 'intensive' farming, machines have to be put in the ponds to mix and aerate the water.

Biotechnological Application

Biotechnology - any technological application that uses biological systems, living organisms, or derivatives thereof, to make or modify products or processes for specific use 1994 UNEP Convention on Biological Diversity - has a wide range of useful applications in fisheries and aquaculture. It creates opportunities, for instance, to increase growth rate in farmed species, boost the nutritional value of aquafeeds, improve fish health, help restore and protect environments, extend the range of aquatic species and improve the management and conservation of wild stocks. Some biotechnologies are simple with a long history of application such as the fertilization of ponds to increase feed availability. Others are more advanced and take advantage of increasing knowledge of molecular biology and genetics, e.g. genetic engineering and DNA disease diagnosis.The field of genetic biotechnology similarly ranges from simple techniques such as hybridization, to more complex processes such as the transfer of specific genes between species to create GMOs (genetically modified organisms).

Over the years, our knowledge of fish breeding requirements has improved and the ability to induce artificial breeding developed through the use of natural or synthetic hormones and/or environmental manipulations. (For example changing photoperiod or water temperature can induce some fish to spawn). These have been key factors facilitating the application of more advanced biotechnologies.

Selective breeding, the maintenance of stocks genetically improved by chromosome manipulation, line crossing, and sex reversal all depend on the controlled breeding of farmed species. These improvements in reproductive technologies have also assisted aquaculturists greatly in their efforts to domesticate aquatic animals. In addition, by making it possible to remove the natural constraints and timing of breeding, farmers are able to mate many more species at times that are most beneficial, and thus helping to ensure a steady and consistent supply of fish for consumption.

Aquaculture Extension

Aquaculture contributes significantly to the rural economy of most of the Asian and other developing countries by providing part- and full-time occupation to the farmers, fishermen and landless agricultural labourers. India and other developing countries of the South Asian region are endowed with ample water resources in the shape of freshwater ponds and tanks for fish culture, but these are not under effective and optimum utilization in spite of highly developed available technologies. Research results have shown excellent production potential as well as economic viability of such technologies, but until these technologies are successfully transferred to the beneficiaries, the desired objective cannot be achieved. Like agriculture, aquaculture is also an agroclimatic-based technology which, when developed in one agroclimatic region, may need modifications and refinement for adoption to another region. Agriculture extension thus involves not only the extension of aquaculture technology, but also certain levels of adoptive research in a particular field environment before it is launched for large-scale extension.

It is a two-way education process in which both scientists and farmers contribute, receive and interact with the involvement of extension workers as a link between the two and a catalyst as well (model). In other words, it is a non-formal adult education programme for educating and training the rural mass to acquire suitable fish farming skills and capabilities with a view to boosting fish production efficiency and the socio-economic condition.

Aquaculture extension is basically an educational process by means of which scientific and technological knowledge of aquaculture is carried to the farmers to upgrade their existing operation and farm management skills. The philosophy behind this process is to change the altitude, enhance the skill and knowledge of the fish farmers to upgrade their aquaculture practice. It also aims at binging maximum possible unutilized and underutilized water areas under modern fish culture operation so as to raise the standard of living of the fish farmers through improving productivity and profitability.

Apart from achieving its own target its overall objective is also to signifiantly contribute towards rural development by improving rural economy, creating additional gainful employment opportunities, fighting malnutrition and preventing rural exodus.

Launching aquaculture tension programme

Any aquaculture extension programme is designed based upon broad national consideration to achieve national goals and targets viz-a-viz local considerations to achieve short-term objectives such as application of composite fish culture in undrainable ponds to improve the aquaculture production level. A local aquaculture extension programme is relatively more definite in terms of scope and target. Like any other aquaculture extension programme, there should be three sequential steps for the dissemination of fish culture technology in undrainable ponds. They are as follows:

- Programme planning
- Programme implementation
- Programme evaluation

Programme Planning

While planning the dissemination of fish culture technology one should always bear in mind that the programme should be a self-regenerating production endeavour and once it is stimulated should continue on its own with a changed attitude and active participation of the recipient. This involves situation-specific strategies. The main components of programme planning are pre-adoption survey of the area, situation analysis, setting programme goals and finally designing strategies in a sequential manner.

Village Survey

Fish culture is basically a rural farming system and hence village survey is the most common method for identification of the difficulties faced by the farmers and to find out the scope and suitability of a specific technology needing to be transferred. The main objective is to get an overall picture of the village and the villagers, their attitude, values, together with their socio-economic conditions and also to locate and assess the available freshwater resources.

It also helps to identify the local institutions, village leaders, progressive farmers, school teachers, village level workers in order to design the most feasible extension strategy and also to establish a permanent rapport to strengthen the extension services.

At micro level it provides information about the socio-economic conditions of individual fish farmers, the pattern of fish farming and the technological gap.

The village selected for the survey should be such that it may represent the locality. Regular contact with important and progressive farmers of the village should be maintained. They should be informed about the objectives of the survey proposed to be undertaken.

Interviews with these persons will provide an overall picture of available natural and human resources and possible areas for development. Finally detailed relevant information may be collected from individual pond owners/fish farmers and fishermen through personal interviews/ questionnaires.

Resource Inventory

- *Availability of water resources:* Various types of water resources are available for fish culture but usually all of them are not fully utilized. Large, medium and smaller types of water bodies are generally available in villages which may be suitable for fish culture. Many small water bodies are found fully shaded by large marginal trees and thereby lying unproductive. Some unconventional types of water areas with potentiality for intensive aquaculture are also available.

  Canal/road and village side small and large ditches, pits emerging due to construction of mud houses etc., are some of the unnoticed and untrapped potential aquaculture resources suitable for seed production and short-term fish rearing.

  Low-lying and swampy areas which are formed naturally due to human activities are also potential sites for undrainable ponds for fish culture.

- *Availability of human resources:* It is a well-known fact that the majority of the people in developing countries live in villages and most of the rural population depend upon agriculture, aquaculture, livestock farming and other allied activities for their livelihood. Human resources are the vital inputs in rural aquaculture development. Rural areas have vast potential of unutilized or underutilized human resources for both men and women, which can be effectively utilized in operating aquaculture.

- *Identification of possible constraints:* A village survey also offers an excellent opportunity to identify various constraints in the background of which an appropriate strategy can be suitably designed.

Financial

Farmers usually do not have surplus funds big enough to be diverted towards reclamation and renovation of existing watersheds as well as construction of new ponds. Initial expenditure for fish culture over fish toxicant, fish seed and supplementary feed is itself a considerably big amount to be exclusively borne by farmers themselves without any credit support. As such, possible sources for mobilizing credit facilities may be identified.

Improper Water Area Distribution Pattern

Like land distribution pattern, major water areas are usually found in the possession of medium and big farmers who bother least about fish culture and concentrate themselves mostly on agriculture, while small and marginal farmers have minimum water holdings at their disposal with adequate manpower potential to be utilized. In some areas most of the water bodies are vested to village institutions, local administrative bodies, etc.

Lack of Technical Knowhow

Several seasonal and perennial ponds without any proper embankments are found lying fallow in a derelict condition due to ignorance and lack of technical knowhow. In some cases farmers fail to follow-up the prescribed package of practices strictly and land themselves in a state of financial turmoil and lose confidence in the viability of newly developed fish farming technologies.

Lack of Stocking Materials and Other Material Inputs

Fish farmers usually face the biggest problem of unavailability of quality fish seed for stocking their pond. Paucity of quality fish seed in the locality force the farmers to stock their ponds without any consideration to proper stocking size, density, species, ratio, etc.

At times, they procure riverine fish seed which is usually mixed with the seeds of predatory and weed fishes. Other material such as fish toxicants are usually localised in its availability. All such problems are also vital for deciding area specific extension strategies.

Marketing Problem

It is a general practice that the fish is sold to middlemen at the pond site who invariably pay lower prices. Due to the perishable nature of the commodity and fear of exploitation by the fish wholesellers, farmers prefer to sell the crop at their pond/farm sites even at relatively lower rates. Information related to marketing practices will add to the scope of the extension programme so that farmers may be educated in marketing management to avoid such exploitation.

Lack of Transport and Efficient Communication System

In remote villages of India and many developing countries where fish culture technology needs to be extended, proper transport and communication facilities are lacking.

Social and Administrative Problems

Ponds remaining unutilized and lying in derelict conditions are common sights in rural areas in spite of a certain level of fish culture know how available with the farmers. In most cases such conditions exist due to family rivalry and non-cooperation among the members of the owners especially when the water areas are under multi-ownership. Poaching and deliberate poisoning of the ponds to destroy the crop are also serious social problems.

In some areas fish culture is supposed to be of a low-caste profession, thus many efficient upper-caste prople remain reluctant to come forward for this venture. Local administration such as Panchayats and Block Level Development Departments are also not always suitable geared enough to ensure rural aquaculture development.

Setting Programme Goals and Planning

In the light of resource inventory and possible limitations suitable target groups may be identified, programme goals may be set up and accordingly suitable extension strategy may be planned. Without such an early insight and planning, the programme may not have firm and realistic footing. Although the fish farmers are the usual target of any fish culture extension programme, all the fish farmers may not be suitable to be involved for immediate participation.

Target groups may be selected on a number of criteria including farming practice, production level, income, education, cultural background, nature, reputation in the society, initiative, liable to change their attitude, etc. Selection of suitable communication channels is also very important. Data collected during the pre-adoption survey provide the necessary information for such selection. After these selections, programme goals may be set up. Goals indicate the direction towards which the programme is oriented. It also provides reference level for evaluating the programme achievements. Examples of goals in such extension programme may be on the following lines:

- Improving the socio-economic uplift of fish farmers and raising the standard of living;
- Bringing 100% of the available undrainable ponds for composite fish culture.

Programme Implementation

"Plan the work and work the plan" is an appropriate term for any extension programme.

Once the plan is laid, all possible efforts should be diverted to ensure that responsibilities are carried out, schedules are followed arid activities accomplished as per the plan. Although the strategies and planning of the aquaculture extension programme are situation specific, some general steps may be cited as follows;

- Through heavy flow of information using mass media, publications, individual and mass contacts, etc., awareness and interest should be created among fish farmers.
- One or two demonstration centres may be set up and the technology of composite fish culture and seed production in undrainable ponds may be demonstrated to maximum possible farmers to let them realize the case of operation, production potential and profitability.
- A set of suitable farmers should be selected initially and be motivated and guided enough so that they strictly follow the different package of practices as per the schedule.
- Proper steps may also be taken to make available the critical material inputs at the pond/farm sites and if the programme permits, subsidy should be given as a token of initial attraction.
- If the availability of quality fish seed is found to be a limiting factor, fish seed may be distributed free of cost or at concessional rate to the farmers at the initial stage. Proper attention may also be paid for extending fish breeding and seed rearing programmes.
- Facilities for proper monitoring of water quality and fish health may be extended through the participation of nearby laboratories.
- Periodical netting for growth check/health inspection should be strictly followed and supervised.
- Self-explanatory/pictorial instruction booklets dealing with basic steps of composite fish culture, control breeding of common carp, techniques of pituitary gland collection, induced breeding of major carps, hatchery operation for carps, nursery and rearing pond managements, techniques of fish seed transport, etc., may be prepared, explained and distributed among farmers.

- Ad hoc training courses should be organized at the demonstration sites on different aspects of fish culture and fish seed production for participating and other interested farmers. Exhibition programme/Fish Farmer's Day should be organized time to time at different places in which live specimens of all the six carp species, other culturable air-breathing fishes, harmful predatory species, weeds, fish feeds, fish toxicants, etc., should be shown and the objective and goal of the programme may be exhibited through models, charts, posters, etc.
- At times a team of a few farmers may be selected on the basis of their leadership quality and performance and sent to visit important aquaculture centres, farms, research institutions, etc.
- Individual contacts through home visits is a very effective extension method. The extension worker must be very clear in this objective during the visit and must do sufficient preparation with regard to subject matter information he is going to deliver to the fish farmer and family members.
- Evening is the most suitable time for organizing an assembly of farmers. Necessary details about practices to be followed the next day may be explained to them during such assemblies. Teaching aids may be used to make the communication effective.

Programme Evaluation

Programme evaluation is the process to determine the extent of success of the executed extension programme in the light of present objectives. It is an important management function in order to ensure effective implementation of the programme. It also helps in the identification of the deficiencies and weakness of the programme so that proper corrective measures may be taken to make it more useful in its future course. Programme evaluation can be conducted once a year or at a specific period of the programme and finally at the concluding phase of the programme. The process of evaluation also depends upon the nature of the programme. A short term and less extensive localized extension programme may be evaluated by the extension workers themselves through the analysis of progress reports, field records, questionnaires, etc. However, broad-based and elaborate extension programmes can be evaluated by specialists in association with the extension workers to determine the effectiveness and impact of the extension programme.

It is convenient to fractionate the whole programme into smaller components for effective and easy evaluation. Fractionation may be done as follows:

- Resources (financial, personnel an material) made available.

- Objective of the programme in clear terms.
- Phases of the programme (evaluation should also be done phase-wise).
- Data collection from records and tabulation.
- Selection of ways, means and methods for the collection of data/ information from participating, non-participating fish farmers, village youths, prominent persons of the locality, etc.
- Sample selection
- Collection of data/information from target and non-target groups
- Tabulation of data
- Data analysis and interpretation of results

To measure the degree of success, certain values have to be associated with the information. Increased fish production level, profit through increased fish yields, knowledge of modern management techniques, fish breeding, fish seed rearing, increased number of ponds/water areas in the area, etc. are some of such measurable values for programme evaluation.

7

Conservation and Utilization of Genetic Resources

New biotechnologies offer not only techniques to improve the conservation of genetic resources, but also new methods to identify, clone, transfer and express genes in different organisms. This has profound implications for the utilization of genetic resources, could broaden the germplasm base from which new genetic combinations can be created, and will allow scientists to pursue their breeding efforts with greater focus and speed.

In vitro conservation and the use of techniques for germplasm exchange of certain plant species are already proving considerably more efficient than conventional methods. Furthermore, the generation of gene libraries provides a valuable adjunct to the conventional germplasm conservation methods. Molecular methods such as Restriction Fragment Length Polymorph (RFLP), isozyme and protein analyses are already being used for the characterization of genetic resources and the identification of useful genes. However, massive use of crop varieties reproduced through in vitro culture, which contain genetically identical copies of the parent, could provoke increasing genetic erosion.

Resistance/tolerance to Stresses

Biotechnologies have been particularly effective in and could have great potential for developing genotypes resistant/tolerant to biotic and abiotic stresses commonly affecting crop, livestock, forest and fish species, and can therefore contribute to reducing inputs of pesticides and to stabilizing agriculture in marginal lands and inhospitable habitats. Genetically engineered de novo synthesis of biopesticides by host species for their own defence, development of biocapsules containing genetically engineered biopesticides and virus coat protein-mediated resistance represent tremendous opportunities for pest management.

It is anticipated that genetically engineered vegetable and cereal crops, resistant to certain insects and viruses or tolerant to herbicides, will be commercially available by 1993. Genetically engineered biopesticides are near approval for sale in some countries. The efficiency of insect-based control of pest arthropods, pathogens and weeds and the production of sterile males for insect control can be enhanced considerably through genetic engineering. Furthermore, biotechnologies are aiding conventional breeding in developing genotypes resistant/tolerant to high and low temperature regimes, drought and excess water conditions and saline and other problem soil conditions. A major concern about these solutions, however, is their sustainability if the improved attributes are based on one or only a few genes, as has often been the case.

Sustainable Production

Biotechnology may be used as a tool in the sustained production of crops, livestock, forest and fisheries by providing opportunities to:

- Develop genetically broad-based varieties/breeds resistant/tolerant to biotic and abiotic stresses;
- Increase efficiency in the conservation and enhanced use of natural resources, including genetic resources and systems;
- Reduce use of off-farm inputs such as pesticides;
- Develop populations adapted to marginal land and problem soil conditions; and
- Produce genetically engineered organisms for degradation of toxic wastes and detoxification of chemical residues on produce or land.

In order to be not only environmentally friendly but also socio-economically viable and attractive, a more vigorous and focused research and development approach, especially by the public sector, is necessary. The private sector should also be encouraged to achieve the desired goal of sustainable development through the use of sound biotechniques.

Management of Genetic Resources

This Background Document provides a summary of recently developed biotechnologies that could be used, or more widely used, in the fishery sector in developing countries. Genetic biotechnologies that can be used in fisheries and in aquaculture include those that help to manage genetic resources and those for genetic improvement. For management of genetic resources, markers can be used in the identification of management units and of endangered species to assist fishery management and they can also help broodstock management in stocking programmes. These markers may be genes, proteins (*i.e.* the products of genes), sequences of DNA or the phenotypic expression of genes (different colours, shapes etc.). In the 1960's, analysis of proteins revealed a wealth of genetic diversity in wild populations. Protein analysis is now relatively fast and inexpensive, but it requires tissue samples to be stored and transported frozen. DNA analysis is becoming the method of choice because of the small amount of tissue needed, the fact that the tissue can be stored dried or in alcohol, and because DNA analysis reveals much more genetic variation than protein analysis. Several kinds of DNA markers exist, such as RFLPs, AFLPs, RAPDs and microsatellites. These, as well as other kinds of markers, can be used to analyse gene frequencies and genetic variation in and between different groups of fish. Studies carried out using these technologies in fish populations have revealed high levels of genetic variation distributed throughout the fish genome.

Genetic improvement technologies

Genetic improvement technologies cover a range of techniques requiring different levels of expertise and resources. Chromosome-set manipulation (*i.e.* polyploidy induction) is an established technique to increase the number of chromosome-sets (ploidy number) in an organism. Temperature, chemical and pressure shocks applied to fish eggs can be used to produce triploid (3 chromosome-sets) individuals that have desirable culture traits. Sex-reversal and the production of single sex groups of fish is also a simple technology that combines hormone treatment and chromosome-set manipulation. Hybridisation, *i.e.* the mating of genetically different groups from the same species (intra-specific hybridisation) or from different species (inter-specific hybridisation), is a simple technique that is now easy to accomplish due to our increased knowledge of reproductive biology.

It can be used to combine good traits from two different species into one group of fish or to transfer a characteristic of one group to another. A problem is that breeding hybrids with hybrids results in a non-uniform and unpredictable group of fish that is generally not well suited for culture. Therefore, for hybrid production, the parent-lines must be maintained pure. The above genetic improvement techniques are considered short-term strategies, where the gains are seen in one or two generations. Selective breeding is a longer-term strategy where gains are accumulated at each generation of selection.

Molecular markers may now increase the efficiency of selective breeding by facilitating the identification of quantitative trait loci (QTLs), *i.e.* genes that control complex characters such as growth rate and environmental tolerance and, secondly, by making it possible to use molecular markers linked to QTLs to identify desirable individuals or families. Genetic engineering and the production of transgenic organisms is an active area of research and development in aquaculture. This is a medium-long term strategy in that development and testing of stable transgenic lines requires time. The large size and hardy nature of many fish eggs allows them to be manipulated rather easily and facilitates gene transfer by direct injection of a foreign gene or by electroporation, where an electric field assists gene transfer. We will briefly discuss currently available biotechnologies in the context of fishery management, aquaculture and conservation respectively.

Marine Conservation

Marine conservation is the protection of marine species and ecosystems in oceans and seas worldwide. It involves protection and restoration of species, populations and habitats and mitigating human activities such as overfishing, habitat destruction, pollution, whaling and other issues that effect marine life and habitats.

Marine Conservation Techniques

Marine conservation work can be done by enforcing and creating laws, such as the Endangered Species Act and Marine Mammal Protection Act. It can also be done by establishing marine protected areas, studying populations through conducting stock assessments and mitigating human activities with the goal of restoring populations.

Marine Conservation Issues

Current and emerging issues in marine conservation include:

- Ocean acidification
- Reducing bycatch in marine fisheries and entangle-ments in fishing gear
- Establishing marine protected areas
- Regulating whaling
- Protecting coral reefs through studying the problem of coral bleaching
- Addressing the worldwide problem of invasive species
- Dealing with the problem of shark finning

Overview

Marine conservation is the study of conserving physical and biological marine resources and ecosystem functions. This is a relatively new discipline.

Marine conservationists rely on a combination of scientific principles derived from marine biology,oceanography, and fisheries science, as well as on human factors such as demand for marine resources and marine law, economics and policy in order to determine how to best protect and conserve marine species and ecosystems. Marine conservation can be seen as subdiscipline of conservation biology.

Coral Reefs

Coral reefs are the epicenter for immense amounts of biodiversity, and are a key player in the survival of an entire ecosystem. They provide various marine animals with food, protection, and shelter which keep generations of species alive Furthermore, coral reefs are an integral part of sustaining human life through serving as a food source (*i.e.*, fish, mollusks, etc.) as well as a marine space for eco-tourism which provides economic benefits. Unfortunately, because of human impact of coral reefs, these ecosystems are becoming increasingly degraded and in need of conservation.

The biggest threats include "overfishing, destructive fishing practices, and sedimentation and pollution from land-based sources." This in conjunction with increased carbon in oceans, coral bleaching, and diseases, there are no pristine reefs anywhere in the world. In fact, up to 88 per cent of coral reefs in Southeast Asia are now threatened, with 50 per cent of those reefs at either "high" or "very high" risk of disappearing which directly effects biodiversity and survival of species dependent on coral.

This is especially harmful to island nations such as Samoa, Indonesia, and the Philippines because many people depend on the coral reef ecosystems to feed their families and to make a living. However, many fisherman are unable to catch as many fish as they used to, so they are increasingly using cyanide and dynamite in fishing, which further degrades the coral reef ecosystem. This perpetuation of bad habits simply leads to the further decline of coral reefs and therefore perpetuating the problem. One solution to stopping this cycle is to educate the local community about why conservation of marine spaces that include coral reefs is important. Once the local communities understand the personal stakes at risk then they will actually fight to preserve the reefs. Conserving coral reefs has many economic, social, and ecological benefits, not only for the people who live on these islands, but for people throughout the world as well.

Human Impact

The deterioration of coral reefs is mainly linked to human activities - 88 per cent of coral reefs are threatened through various reasons, including excessive amounts of CO_2 (Carbon Dioxide) emissions. Oceans absorb approximately 1/3 of the CO_2 produced by humans, which has detrimental effects on the marine environment. The increasing levels of CO_2 in oceans change the seawater chemistry by decreasing the level of pH.

This process is also known as acidification. Acidification negatively affects the carbonate buffering system and drops the carbonate saturation by 30 per cent, which results in a decrease in reef calcification. Reductions in calcification have negative implications on calcifiers, such as corals and shellfish. Some examples include diminishing coral resilience from bleaching, decreasing organisms' ability to fight off predators, inhibiting their potential to compete for food, and altering behaviour patterns.

When the bottom of the food web declines tremendously due to acidification, the food web and the whole marine conservation effort is jeopardised. Although humans cause the greatest threat to our marine environment, humans also have the ability to create effective management plans that will be the key to successful marine conservation. Although the most widely known conservation tool is the MPA, one of the best marine conservation tools simply stems from smarter individualist choices we make in efforts to reduce CO_2 emissions on a daily basis.

Techniques

Strategies and techniques for marine conservation tend to combine theoretical disciplines, such as population biology, with practical conservation strategies, such as setting up protected areas, as with Marine Protected Areas (MPAs) or Voluntary Marine Conservation Areas. Other techniques include developing sustainable fisheries and restoring the populations of endangered species through artificial means. Another focus of conservationists is on curtailing human activities that are detrimental to either marine ecosystems or species through policy, techniques such as fishing quotas, like those set up by the Northwest Atlantic Fisheries Organisation, or laws such as those listed below.

Recognising the economics involved in human use of marine ecosystems is key, as is education of the public about conservation issues. This includes educating tourists that come to an area that might not be familiar of certain rules and regulations regarding the marine habitat. One example of this is a project called Green Fins that uses the SCUBA diving industry to educate the public based in SE Asia. This project, implemented by UNEP, encourages scuba diving operators to educate the public they teach to dive about the importance of marine conservation and encourage them to dive in an environmentally friendly manner that does not damage coral reefs or associated marine ecosystems.

Technology and Halfway Technology

Marine conservation technologies are devices used to protect endangered and threatened marine organisms and/or habitat. Marine conservation technologies are innovative and revolutionary because they reduce bycatch, increase the survivorship and health of marine life and habitat, and benefit fishermen who depend on the resources for profit. Examples of technologies include Marine Protected Areas (MPAs), Turtle

Excluder Devices (TEDs), and Radio-Frequency Identification (RFID). Commercial practicality plays in important role in the success of marine conservation because it is necessary to cater to the needs of fishermen while also protecting marine life.

Turtle Excluder Devices for example (TEDs) remove a major threat to turtles in their marine environment. Many sea turtles are accidentally captured, injured or killed by fishing. In response to this threat the National Oceanic and Atmospheric Administration (NOAA)worked with the shrimp trawling industry to create the TEDs devices.

By working with the industry they insured the commercial viability of the devices. Basically, a TED is a series of bars that is placed at the top or bottom of a trawl net, fitting the bars into the "neck" of the shrimp trawl and acting as a filter to ensure that only small animals may pass through.

The shrimp will be caught but larger animals such as marine turtles that become caught by the trawler will be rejected by the filter function of the bars. Similarly, halfway technologies work to increase the population of marine organisms, however, it does so without behavioural changes and "addresses the symptoms but not the cause of the declines". Examples of halfway technologies would include hatcheries and fish ladders.

Laws and Treaties

International laws and treaties related to marine conservation include the 1966 Convention on Fishing and Conservation of Living Resources of the High Seas. United Stateslaws related to marine conservation include the 1972 Marine Mammal Protection Act, as well as the 1972 Marine Protection, Research and Sanctuaries Act which established the National Marine Sanctuaries programme.

In 2010, the Scottish Parliament enacted new legislation for the protection of marine life with the Marine (Scotland) Act 2010. The provisions in the Act include: Marine planning, Marine licensing, marine conservation, seal conservation, and enforcement.

Organisations and Education

There are marine conservation organisations throughout the world that focus on funding conservation efforts, educating the public and stakeholders, and lobbying for conservation law and policy. Examples of these organisations are Oceana (non-profit group), the Marine Conservation Institute (United States), Blue Frontier Campaign (United States), Sea Shepherd Conservation Society, Frontier (the Society for Environmental Exploration) (United Kingdom), Marine Conservation Society (United Kingdom), Community Centred Conservation (C3), The

Reef-World Foundation (United Kingdom) and Australian Marine Conservation Society.

On a regional level, PERSGA- the Regional Organisation for the Conservation of the Environment of the Red Sea and the Gulf of Aden, is a regional entity serves as the secretariat for the Jeddah Convention-1982, one of the first regional marine agreements. PERSGA Member States are: Djibouti, Egypt, Jordan, Saudi Arabia, Somalia, Sudan and Yemen.

Approaches the genetics of behaviour

We describe in order of increasing complexity how a researcher might approach the genetics of a behaviour of interest. For each method, I describe its aim, the general approach, its strength and weaknesses and the kinds of traits and animals for which it is most suited.

'Low tech' approaches

The first step in approaching the genetics of a behavioral trait is to determine if the behaviour is repeatable, which involves measuring the same behaviour on the same individual several times. Repeatability is the proportion of phenotypic variance attributable to the individual, which could be caused by additive genetic variance or environmental variance with long-lasting effects. Evidence that a behaviour is repeatable indicates that it might have a heritable component and is amenable for further genetic dissection.

The repeatability of a trait sets an 'upper bound' to heritability. However, it is worth considering that nongenetic effects can also produce stable behaviour; repeated reinforcement including learning can also lead to stability. Another indication that there might be a heritable component to the behaviour is if it is consistent across contexts; in this case, the stability occurs via a correlation between individual behaviours in different contexts, or a behavioral syndrome.

Another relatively low-tech approach to the genetics of a behaviour is to ask if it differs across populations that occur in different kinds of environments. If so, then population differences in the behaviour could reflect an evolved response to differing selective pressures. Alternatively, the differences could reflect environmentally-induced changes within individuals. A way to disentangle these two hypotheses is to rear animals from the different populations in a 'common garden', or in the same environment. If population differences in the behaviour are preserved under common environmental conditions, then the difference between populations might be genetic in origin. A common garden experiment represents the first step towards approaching the genetics of complex traits such as behaviour, but there are at least two important factors to consider before concluding that population differences are genetic in origin. The first is that F1s reared in a common garden could still experience parental effects from their wild-caught parents.

To control for such environmental effects, it is preferable to rear the offspring of lab-born individuals and to compare the phenotype of the F2s. Second, GxE interactions should also be considered when populations do not differ in a common environment. For example, individuals from both populations might converge in their response to the common garden environment, but the extent of response might have a genetic component. A reciprocal transplant experiment might reveal genetic differences that are not apparent in a common environment.

An advantage to using population differences as a clue to which behaviours are likely to be heritable is that traits that differ across populations in different environments are probably linked to fitness. Several common garden studies on fishes have shown that population differences in behaviour are often preserved in a common garden. For example, Lahti *et al.* 2001 found evidence for a genetic basis to aggressive behaviour in brown trout by comparing different types of populations. The authors reared fish from 10 populations in a common environment. The populations can be grouped into the following types: resident (nonmigratory), sea-run (migratory) or lake-run (migratory), so there were replicate populations for each type. Contrary to the expectation that resident trout are more aggressive than migratory forms, they found that sea-run populations were consistently more overtly aggressive towards opponents than the other types of populations.

There is a rich literature comparing domesticated and wild populations of salmonids, which has revealed genetic influences on several types of behaviour (reviewed in). When reared in a common garden, hatchery fish show reduced antipredator responses are more bold towards a novel object and are frequently more aggressive. However, wild Atlantic salmon (*Salmo salar*) became dominant over hatchery salmon if they were given an opportunity to establish residence, suggesting that the expression of genetic differences between hatchery and wild salmon depends on the environmental context.

Quantitative Genetic Approaches

There is a long history of examining the genetic basis of continuously distributed traits using quantitative genetic techniques, which are based on the phenotypic resemblance among relatives due to shared genes. Unlike some of the reverse methods, quantitative genetic techniques measure phenotypes, not genes, and rely on the resemblance among relatives to infer genetic effects. Quantitative genetic approaches assume that many genes of small effect influence the phenotype, thereby producing a continuous distribution (however, recent QTL studies have called this assumption into question). By measuring the phenotype on individuals of known relatedness, we can partition the phenotypic variation into environmental, genetic and gene x environment variance components.

The simplest quantitative genetic approach is to ask whether the trait of interest runs in families, which would suggest that there might be a genetic component to it.

However, families often share a common environment so cross fostering experiments are useful here. Like repeatability, a demonstration of full-sib resemblance sets an upper limit to the heritability. More complicated breeding designs involve estimating the resemblance of parents and offspring using regression, or generating full and half-sib families to disentangle parental effects from genetic effects using ANOVA-based approaches, which are thoroughly described elsewhere. These approaches estimate heritabilities, genetic correlations, parental effects and non-additive genetic variance.

The most powerful quantitative genetic experiments estimate the G matrix. The G matrix refers to the multivariate matrix of genetic variances and covariances between several different traits. Although laborious to quantify, obtaining a G matrix allows us to predict the consequences of selection on any given trait, and, with a few assumptions about the stability of the G matrix through time, allows us to retrospectively analyze selection in the past. In addition to satisfying the assumption of stability required for retrospective selection analysis, it is also desirable to compare G matrices to determine whether genetic constraints are limiting. If genetic correlations can be uncoupled, then the relationship between traits might respond to selection, and the G matrix will, itself, evolve. Alternatively, genetic correlations might limit the number of different configuration of traits that are possible, which would be reflected in an invariant G matrix through time. Fortunately, there are several different techniques available to compare the structure of G matrices but there is debate about the best method.

A relatively new powerful technique known as the animal model allows us to estimate genetic variances and covariances without performing breeding experiments in the lab. Therefore this method is particularly useful for field studies.

The animal model is powerful because it takes advantage of all types of resemblance between relatives to partition variance components; it uses all the available information about relatedness from a pedigree. Moreover, this method is suitable for unbalanced datasets, and can accommodate missing values.

Another approach for estimating genetic variances and covariances is to perform an artificial selection experiment, which involves selective breeding of individuals.

Selection experiments have shown that many behavioral traits can respond to selection. Artificial selection experiments are only suitable for organisms that can be kept in the lab, and which ideally have a short generation time. Selection experiments should only be undertaken when different selected lines can be replicated and when sample sizes are sufficient to control for drift and inbreeding. The pitfalls and perils of selection experiments are reviewed in.

Artificial selection experiments can be particularly insightful when they vary the environmental context in which selection occurs. For example, in a very interesting series of selection experiments on medaka (*Oryzias latipes*, Ruzzante and Doyle (1991, 1993) showed that the correlated response of aggressiveness to selection for fast growth depended on the ecological context in which the selection took place. In these experiments, Ruzzante and Doyle selected for fast growth under two different conditions: when food was clumped and could be defended, and when food was dispersed.

The same limited amount of food was added to each treatment. Selection for fast growth produced a correlated response to selection on aggressiveness, but only when food was defensible. Interestingly, levels of aggressiveness *decreased* in the fast-growth lines, which they interpret as reflecting indirect selection for 'social tolerance'.

One of the strengths of a selection experiment is that it can tell us whether there are packages of traits that are linked together and respond to selection in concert. For example, selection for increased and decreased stress responsiveness, as measured by the change in circulating concentrations of cortisol in response to handling stress, produced several correlated effects on physiological and behavioral measures in rainbow trout.

Relative to trout that did not release very much cortisol in response to handling stress, 'high responding' trout were more aggressive in a new environment were more active in the presence of an intruder and took longer to acclimate to a new environment. Moreover, these behavioral differences were also accompanied by changes in the brain monoaminergic systems.

These results suggest that an entire suite of physiological and behavioral changes accompanied modifications to stress responsiveness. Interestingly, the packages of physiological and behavioral traits that changed together are analogous to different 'coping styles' that have been identified in other vertebrates.

Candidate Gene Approaches

Over the past few decades, there has been increasing evidence that the molecular functions of many genes are highly conserved across species. For example, studies on transgenics have revealed that genes from one species accomplish similar functions in distantly-related species. The incredible conservation of gene function in living things allows us to apply genetic information about other species to the organism of interest. This means that we can apply the genetic information gained from studies on model organisms to nontraditional models for which genomic information is not available.

Genes could become candidates based on genetic studies in other species (*i.e.* when polymorphism has already been identified and associated with behavioral variation, Fitzpatrick *et al.*, 2005), or because a physiological pathway leading to behaviour of interest is well-understood, suggesting hypotheses about which genes to examine. For example, the neuroendocrine mechanisms underlying circadian rhythms are well described for several mammalian species. Therefore, we can ask whether the expression or structure of genes along those pathways differ among individuals or among groups. Candidate gene expression approaches typically involve measuring the abundance of mRNA transcript in a particular tissue (usually brain, in the case of behaviour) using quantitative real-time PCR or Northern blots.

One of the attractions of the candidate gene approach is that detailed genomic information on the species is not necessary, therefore this technique can be applied to non-model organisms by designing degenerate primers based on other species. Another attraction of the candidate gene approach is that unlike quantitative genetic approaches, it does not require complicated breeding designs or data on the relationships among individuals – comparisons can be made across individuals of unknown relatedness because the genes are being measured directly. Therefore this approach is suitable for organisms that do not readily breed in the lab.

Summaries of using the candidate gene approach for studying behaviour are well-described elsewhere. Notable examples include work on the *for* gene and foraging behaviour, and the 'fruitless' gene and mating behaviour in Drosophila and the promoter region of the vasopressin gene and parental behaviour. The candidate gene approach for studying the genetics of fish behaviour is a relatively unexplored but promising future research direction.

While the candidate gene approach is very attractive and has great appeal especially for nonmodel organisms, there are some drawbacks. First, candidate gene studies are biased; what if the ultimate source of genetic differences between groups lies further up or downstream of the particular candidate gene? One approach to this issue is to choose a pathway that is probably associated with the behaviour and look at gene expression at several points in the pathway. Second, candidate gene studies are purely correlative. Concluding that the gene is really associated with the behaviour requires further experimentation. Perhaps most seriously, candidate gene approaches work best when the trait is influenced by just few genes of major effect, which is probably the exception rather than the rule for behavioral traits.

Genomics

'Genomics' refers to the study of the structure, content and evolution of genomes, including the analysis of the expression and function of genes and proteins. What distinguishes genomics from other branches of genetics is that it looks at the whole genome simultaneously, rather than focusing on one gene at a time. Luckily, doing genomics does not require a full genome sequence, and there are already several good examples of applying whole-genome approaches to nonmodel organisms and ecologically-relevant traits, including behaviour. Genomics has been hailed as an opportunity to integrate mechanistic and evolutionary approaches to studying behaviour because information about genes provides neuroscientists and behavioral ecologists with a 'common language'.

8

Biotechnology in Fisheries and Aquaculture

The fisheries and aquaculture industries in Asia contribute about 45 percent of world fish production . These industries are significant contributors to the food supply, livelihood, foreign exchange earnings and socio-economic stability in rural areas of several Asian countries. At the global level fish production is stagnant, at around 100 million tonnes, or has even slightly declined in recent years. Overharvesting of the world's oceans is recognized as a global threat. To obviate this threat, the production of cultured fish must be accelerated. The ability to produce transgenic fish and shellfish in culture, which grow faster and larger with more efficient utilization of nutrients, is of particular value to developing countries, not only as a source of food, but also as export products.

Biotechnology offers great promises for fish disease control and feed production as well. Gynogenetic and polyploid progeny production and the use of growth hormones have significantly contributed to fish production . The application of hypophysation (injecting of pituitary gland extracts into mature fishes for the induction of spawning) has been widely practiced in Asia during the past two decades. The use of purified gonadotrophin hormone synthetic luteinizing hormone-releasing hormone (LH-RH) has further improved induced spawning techniques.

In China, following detailed discussions and workshops in the 1980s, a policy and plan for aquaculture biotechnology were included in the Seventh Five-Year Plan (1985-90). In 1987, they were also listed in the "863 Programmes", the state development programmes on high technologies, and received great attention from the Ministry of Aquaculture. Fish cell culture, sex control, chromosome manipulation, and to some extent transgenic production, have been the main areas of work.

China's success in fish cell culture, of grass carp, was reported in 1979 by the Freshwater Fishery Research Institute of Zhejiang Province. Up to now, more than ten fish cell strains have been cultured for producing vaccines. Bioreactors for industrial fish cell culture are being developed. Seaweed cell culture in China was first reported in 1985 and is being developed for further breeding work. Diploid gynogenesis, coupled with sex control is being used for mass production of all female common and crucian carps. Although triploid hybrids between 4n common carp (male) and 2n grass carp (female) have been obtained, the work is confined to laboratories.

China was one of the pioneers in transplanting of fish cell nucleus. In the mid-1970s, the reciprocal transplant of nucleus was accomplished between common carp and crucian carp. Of the cell hybrid fishes, 30 grew to sexual maturity and produced F2 and F3 progenies. In 1984, the F3 progenies were mass produced and these showed stable heredity and had 30 percent faster growth than their grandparents.

A similar practice was carried out on grass carp and Megalobrama amblycephala; 26 of the hybrids grew well, matured sexually and were mass produced by aquaculturists. In 1988, live fries were produced from the fusion of blastulla cells and eggs in different fishes:

Paramisgrunus dacryanus, Megalobrama amblycephala, grass carp, common carp and red crucian carp. This technique holds great promise for fish breeding.

In fishes, the two sexes differ in their productivity and body weight. For instance, in Tilapia males grow faster than the females, the converse of which is true for carps. Of the several methods of sex manipulation, through distant hybrids, hormone treatment, induced diploid gynogenesis, induced triploids, antoimmure castration, surgical castration, and gene technique, in China the first two are commonly used. Hybrid populations between Tilapia nilotica (female) and T. aurea (male) were 95 to 98 percent male, and on an average yielded about 25 percent higher than the higher yielding parent (T. nilotica). Moreover, the hybrid fish could tolerate lower water temperatures, and could be harvested more easily. Such all-male populations are already being mass cultured in China, the Philippines, Thailand and other countries. Israel has made comprehensive use of this technique.

As regards the use of hormones, using steroid hormones, induced females of T. mossambicais when crossed with male T. nilotica, produced super males with a "yy" sex chromosome combination. The "super males" crossed with female T. nilotica produced 100 percent male progenies which possessed the twin superiorities of being hybrid as well as male. The hybrids grew 38.5 percent faster than T. nilotica and the mass production increased by 43.4 percent. Although a sizeable gain, the difficulty in identifying the "super male" fish at fry stage is a major constraint to widespread exploitation.

As regards genetic engineering, Chinese scientists have succeeded in identifying and isolating the gene controlling growth hormone (GH) from salmon, common carp, grass carp and silver carp. Transgenic carps carrying human growth hormone gene have been produced. Antifreeze gene has also been identified and used for improving the antifreeze ability of Tilapia.

Successful cryopreservation of fish sperm, involving eight species has been established in China.

India, with a coastline of 7 512 km has 2.02 million km^2 of Exclusive Economic Zones, 4.7 million hectares of identified freshwater areas, and 1.7 million hectares of brackish-water areas. Its annual production of fish is around 4 million tonnes, half of it from aquaculture, and there is ample scope for enhancing fish production in the country. To supplement the fish feed resources, the use of bacterial cells as larval fish feed and processed aquatic plants as grow-out feed has been tested with success in India .

In Japan, chromosomal manipulation and modern biotechnology are being widely applied to improved fish production. Gynogenesis, artificial sex reversal and induced polyploidy are being routinely used for increasing productivity as well as the quality of fish.

The growth hormone of salmon and yellowtail is being mass produced through transgenic bacteria and used as a growth promoter on other fish such as rainbow trout. In the field of transgenic fish, the introduction and manifestation of the crystalline genes of chickens into killifish embryos were successfully achieved in 1986 at Kyoto University . Other useful genes are being identified and introduced into commonly used fish species.

New vaccines are also being developed to treat fish diseases. As regards fish feed, the development of microparticulate diets for larval fish to replace live foods in hatcheries is a major achievement. Mass production of ayu and carp fry in Japan with such diet formulates has been successful.

Malaysia has a relatively long coastline, stretching for about 4 334 km and its marine resources contribute about 9.5 percent of the total value added in the agricultural sector; production has been increasing at a high annual growth rate of about 8 percent over the past decade.

Moreover, the industry employs about 120 000 fishermen. The country clearly sees the scope for the application of biotechnology in fish production through genetic improvement of cultured fish, disease resistance and improved growth rate. Yet there is a gap in absorption or diffusion of the proven technologies.

Besides fish, marine algae are now recognized as important resources for the production of valuable chemicals in Malaysia and local species of marine macroalgae (seaweed) are being screened for valuable chemicals. Some of the species studied and their products include: Gracilaria changii for agar and Turbinaria conoides, Sargassum baccularia and Sargassum siliquosum for algiric acid and antimicrobial compounds.

Singapore considers its marine bioresources as one of the most important and the government is highly supportive of the marine biotechnology venture. The National University of Singapore (NUS) is playing a leading role in this direction. Its Bioprocessing Technology Unit, established in 1990, concentrates on two main areas: fermentation and cell culture; and the isolation, separation and production purification technology.

In addition, the Unit also supplies to specific users, especially synthesized antibodies for immunobiological assays and analyses, and custom-made digonucleotide primers for nucleic acid research. In 1992, the NUS and the Economic Development Board (EDB) established a Bioscience Centre in the University, with the following objectives: (i) to provide a multiuser central facility for biotechnology, (ii) to serve as a vehicle for collaborating with industries and other institutions in Singapore or abroad, in areas of common interest, and (iii) to provide training in biotechnology.

The areas of research of the Centre are: bioactive compounds from marine organisms; toxins and venoms research; plant bioactive products; and products from microorganisms. Other related new establishments are the Centre for Natural Product Research and the Centre for Medical Informatics. The Reef Ecology Study Team of the Department of Zoology in NUS is actively surveying the distribution of the various reef organisms and the Bioscience Centre is screening organisms for antifouling, antibacterial, antitumour, antiviral, antifungal, anti-inflammatory and insecticidal compounds. Antibacterial compounds are being isolated and characterized from horseshoe crabs. Biological properties and potential therapeutic application of toxins and other bioactive compounds in two marine species, namely the "Thunder Crab" (Lyphozozymus pictor) and the stonefish (Sysanceja horrida), have been studied.

In addition, a few bioactive compounds have been purified from five local sea anemone species. These include a cytolysin, a trypsin inhibitor and two putative neurotoxins. Transgenics are also being attempted in a few fish species such as Marble Goby, grouper and salmon. The Marble Goby growth hormone gene has been cloned. NUS scientists are now in the process of combining this gene with a heterologous promoter to make a gene construct for transgenesis. Aquaculturists have been advised of the applications of (i) thyroid hormones in enhancing embryo and fry development; and (ii) gonado-tropin releasing hormones to help induce yolk formation and spawning. The hormones are being encapsulated into synthetic capsules for commercial use. Immunologists are studying the immune responses in fish and are developing various disease control methods, including vaccination with attenuated disease agents to develop antibodies.

Singapore is one of the major exporters of tropical aquarium fishes in the world. The DAN fingerprinting technology is being used to identify proper breeder stocks to be used in breeding programmes to maintain the quality and health of the fishes.

Aquaculture is most developed in Thailand. A Marine Biotechnology Unit (MBU) was established jointly by the National Centre for Genetic Engineering and Biotechnology and Aquatic Resources Research Institute of Chulalong Korn University in 1987. The Unit has been screening algae strains for feed purposes. Using cloned mitochondrial DNA, RFLP patterns of giant freshwater prawns (Macrobrachium rosenberghi) were studied for analysing variation patterns to design appropriate breeding plans. Studies are in progress to manipulate the Gonad Inhibiting Hormones (GIH) for regulating the growth, maturity and reproductive success of prawns.

To control the most common and yield-reducing disease, vibriosis, caused by the bacteria Vibrio anguillarum, of the widely cultured fish seabass, vaccines have been developed. Chromosome manipulation of oyster has also been standardized. Studies on the gametogenic cycle of Thai abalone (Haliotis ovina) are proving helpful for resource regulation and management, and for broadstock conditioning, induction of spawning, control of mating and seed production for land-based production. An outdoor mass culture of benthic diatom, Navicula sp., as a food source of abalone spot was also standardized recently.

Research on marine natural products in Thailand began a few years ago and provides valuable information on the structures and potential applications of the compounds and their derivatives from Thai sponges. Such information could be, apart from the bioactivities of the compounds that can possibly be applied further to pharmaceutical use, helpful in the chemotaxonomy of Thai sponges. The use of seaweeds as a source of phycocolloids is being promoted. The Biopolymer Research Unit (BU) of the Srinakarinwirot University, established in 1987, has developed a simple and low-cost technique for agar extraction from Gracelaria sp. seaweeds which could be applied at the village level. By developing and using chitin filter medium, the technique has improved the quality of the agar produced. The chitin could also be changed to chitosan, another valued product that is biodegradable and nonpolluting and can be used as a binder in recycling paper. This technology changes the shrimp wastes into more value-added products. The Unit is currently operating a project entitled "Technology Development of Shrimp Shell Processing". Other work relates to the development of biomembranes, copolymer membrane and biobeads. Another value-added product is a fluorescence pigment of R-phycoerythrin for protein markers.

The Marine Biotechnology Centre at Burapha University is concentrating on population dynamic studies of phytoplankton and zooplankton and identification of new pharmaceuticals from marine organisms. From the various species screened, the Centre found that 21 marine species had shown some degree of antimicrobial action.

Twenty-three extracts from seven species had shown positive results when subjected to haemolytic testing against erythrocytes of 21 species. From the above, it can be seen that the following biotechnology-related research work on marine organisms is in progress: studies on bioactive metabolites from Thai sponges, isolation of pure culture of some marine microalgae, isolation of pure culture of some marine bacteria, and the use of microalgae in biological control of marine animal diseases.

While considerable progress has been made in fish biotechnology in Asia, the following aspects need to be further strengthened :

- Genetic improvement of cultured fish for improved growth, disease resistance and adaptability to new farming systems.
- Development of low-cost but efficient grow-out diets for cultured fishes using locally available feedstuffs and microbiological processes.
- Utilization of detailed and microbial feed sources for pond cultured fish.
- Production of vaccines and monoclonal antibodies for the prevention and control of bacterial and virus diseases.
- Development of drugs and other biotechnological products from marine organisms.
- Improved availability of mother stocks and seeds, especially of brackish water marine fish and shellfish.

Cytogenetic and FISH analysis

The differential diagnosis between 'true' SAA and hypoplastic MDS is often difficult because of a reduced bone marrow cellularity.1,2 Fundamental discriminant clues can be provided by trephine biopsy and by the identification of a chromosome aberration,3 that, however, has been reported in SAA patients too.

We report on five patients, aged 20 59 years, observed at our Institution between March 1989 and August 1995. Their main clinico-hematological characteristics. Cytogenetics and FISH results are report. At diagnosis an initial cytogenetic analysis showed a normal chromosome pattern in all the patients. All of them evolved into AML showing clonal karyotype defects described according to ISCN4 (trisomies and monosomies were considered clonal when present in at least two and three metaphases, respectively). We decided to employ FISH retrospectively in order to establish if any minor subclone, carrying the chromosome abnormality detected by cytogenetics at the time of AML transformation, was already present at diagnosis. In order to reinforce the interphase FISH analysis of small subpopulations of cells we decided to apply the probe discovering the clonal chromosome defect along with an additional probe used as an internal control (dual-colour FISH). All probes were commercially obtained.

The digoxygenated 5q31 DNA probe was applied to metaphase cells only along with the 5 probe. Its cut-off value was obtained by screening 1000 mitoses from 10 normal controls. We determined the one-sided 95% confidence limit using a binomial distribution for the proportion of metaphase cells that were normal or that carried the del(5)(q13q31). We detected the probability of discovering 0, 1, 2, 3 cells with del(5)(q13q31) by using the upper boundary of the 95% confidence interval.5 As in normal subjects, 3 was the highest number of cells lacking the 5q31DNA probe signal, probably because of a hybridization failure, the normal range for the cells with this karyotypic defect was fixed at 0.8%.

Monosomy 7 or trisomy 8 or other numerical defects were considered clonal by metaphase and interphase FISH when the percentages of the cells with such abnormalities were above the cut-off value determined for each numerical change. The cut-off value was obtained by screening 25 mitoses and 200 interphase cells from 10 normal controls. Monosomy 7 cut-off was determined by applying the 7 probe together with a 12 probe. We calculated the mean percentage of mitoses or cells showing two spots due to the internal control and only one spot due to the 7 probe plus three times the standard deviation. Monosomy 7 cut-off value for metaphase FISH was 8%; that for interphase FISH was 5%. Trisomy 4 and 8 cut-offs were obtained in the same way and were fixed at 0% for both the numerical defects when considering mitoses, at 2.3% and at 2.5%, respectively, when examining interphase cells.

Despite the high interphase FISH cut-offs for monosomies and trisomies making the identification of minor subclones marked by such abnormalities difficult, this technique succeeded in demonstrating that the clonal defect was already present at disease onset in a few interphase cells obtained from four of our patients. This result was confirmed by metaphase FISH, that detected similar percentages of clonogenic cells in the same patients, suggesting that, initially, the chromosome defect was really present in a small subpopulation of cells. Moreover, a careful review of the karyotypes, obtained at disease onset and considered normal on an initial conventional cytogenetics investigation, demonstrated that few mitoses of the same four patients carried the clonal abnormalities observed on leukemic transformation and retrospectively identified by FISH. These cytogenetics and FISH results might be explained by the very small size of the abnormal clone and not by the low mitotic rate with poor quality metaphases often observed in patients with hypoplastic marrows. In both situations, however, FISH plays an important role in identifying clonogenic cells that may easily escape cytogenetic detection.

An intriguing point in patients with hypocellular marrow is the differential diagnosis between 'true' SAA and hypoplastic MDS.1,2,3 In our patients FISH alone was unable to discriminate between true SAA and hypoplastic MDS, as the presence of a chromosome abnormality per se does not help to make a discriminatory diagnosis. However, karyotypic defects such as monosomy 7, trisomy 8 and sporadically trisomy 6 have been observed in SAA evolving into MDS or into AML.3 Therefore, an overlap between SAA and MDS may actually exist.2 This possibility is further illustrated by our patient No. 5, who showed an abnormality identifying a particular subgroup of MDS with a very low probability of evolving into AML, ie a del(5)(q13q31), accompanied by an increased marrow reticulin on trephine biopsy, a sign predictive of AML evolution.1 This case initially classified as hypo-MDS ultimately developed into AML-M4 88 months after initial diagnosis.

In conclusion, the present FISH study was made possible by conventional cytogenetics, the only reason that the abnormalities present in each case were known. Cytogenetics and FISH were complementary; the latter succeeded in discovering minor subclones at diagnosis and, demonstrating their expansion, it was able to predict AML transformation.

A Population Genetics Model

Now that you have seen what happens in a situation of no evolution (Hardy-Weinberg Equilibrium), let's develop a model of natural selection. Remember that natural selection occurs if genotypes don't have equal fitness. We will model a situation in which natural selection is the ONLY form of evolution occurring (there will be no mutation, genetic drift, non-random mating, or gene flow--remember, these are the other forms of evolution, as introduced in the last lecture.) So for this model, we will assume (as we did for Hardy-Weinberg Equilibrium) infinite population size, no movement of individuals among populations, no mutation, and random mating. However, we will see what happens when the different genotypes for the trait differ in fitness. First, consider the fitnesses of genotypes. Remember that fitness is a measure of how well individuals with a certain genetic trait are expected to survive and reproduce. We will still consider a single gene with two alleles (A and a) and three genotypes (AA, Aa, and aa.) Natural selection will occur if individuals with different genotypes differ in fitness. So to understand and model natural selection, a first important point is to determine how to measure fitness.

Remember that natural selection occurs because individuals with some traits (those with higher fitness) survive and therefore reproduce more than do other individuals. The important feature here is reproduction--it is those traits that are reproduced most that necessarily become more common from generation to generation. So we will measure fitness based on the degree to which traits (in this case, different genotypes) reproduce. A first measure of the fitness of a genotype is absolute fitness: the number of offspring an individual with a particular genotype has on average. Absolute fitness could be studied in natural populations; one could determine that, for example, individuals with a certain genotype have on average six offspring. A problem with this measure is that there is no way to determine, just looking at this number, whether having an average of six offspring means that this form has high or low fitness. If AA individuals have six offspring, Aa have three offspring, and aa have one offspring, then six is the best, and we would say it represents high fitness. On the other hand, if AA individuals have six offspring, Aa have twelve, and aa have thirty, then we would say six represents very low fitness. The important point to note here is that whether a particular absolute fitness is high or low depends on the fitnesses of other genotypes in the population. So we will develop a second measure of fitness that accounts for this.

A second measure of the fitness of a genotype is relative fitness: the degree to which individuals with a particular genotype reproduce in proportion to individuals with other genotypes. Typically, relative fitness is measured with respect to the genotype that has the highest fitness. The genotype with highest fitness has a relative fitness value of one; the other fitnesses have relative fitness values that are fractions of one, telling us how well these individuals reproduce with respect to the individuals of the genotype that reproduces best. At this point, you should have all the information you need to work all the problems in Chapter VI of your lab manual, and you should try to start working on them. Here are some tips on how to work the problems: Your first step should be to determine whether you're working a problem for a population in Hardy-Weinberg Equilibrium or a problem in which natural selection is occurring. Remember that the difference between the two models is in whether or not different genotypes have different fitness. If the genotypes all have the same fitness (all survive and reproduce equally well) that should tell you you're looking at a Hardy-Weinberg Equilibrium problem. If the genotypes differ in fitness (if they don't survive and reproduce equally well) that should tell you you're looking at a natural selection problem.

If you have a Hardy-Weinberg Equilibrium problem there are two places the problem could start. You could be given allele frequencies to start out or you could be given genotype frequencies to start out. So your next step in solving the problem is simply to figure out, and write down, which you've got: a llele frequencies or genotype frequencies.

In a Hardy-Weinberg Equilibrium problem remember that if you have an allele frequency, you can figure out the other allele frequency (since p+q=1) and you can figure out any genotype frequency (Freq(AA)=p2, Freq(Aa)=2pq, Freq(aa)=q2.) So if you're given an allele frequency, use it, and get whatever information the question asks for. If you're given a genotype frequency rather than an allele frequency, your first step should be to calculate an allele frequency, because once you have the allele frequencies you can figure out anything else the question might require. To calculate an allele frequency from a genotype frequency, you need the genotype frequency of a homozygote (that is, either AA or aa.) Once you have that, take the square root to get the allele frequency of the allele that makes up that genotype. If you have a natural selection problem then there are three places the problem could start, and you need to figure out which one you have. As with Hardy-Weinberg problems, things go more smoothly if you have allele frequencies, so your first step is to get an allele frequency (if you're not already given one.) Here are the three places a natural selection problem could start, and what your first step should be for each possible starting place (once you've completed the first step, you should be able to use the natural selection formulas to work the rest of the problem--it is often the first step that is hardest.)

1. You could be given an allele frequency in gametes at the start of a generation. If you are, use the allele frequency to determine the other allele frequency (remember p+q=1) and use the fitness information you are given (be sure you take this and calculate relative fitnesses if you're not already given them) and use the formulas you have to get the information required by the question.
2. You could be given a genotype frequency in zygotes. If you are, remember that genotype frequencies in zygotes are p2, 2pq, and q2, just like in Hardy-Weinberg Equilibrium. If you have the genotype frequency of a homozygote in zygotes, take the square root to get the allele frequency that was present at the start of the generation, and proceed as you did in the first case where you were given the allele frequency directly.

You could be given a genotype frequency in adults. If you are, then you are in a situation where the population is NOT in the proportions of Hardy-Weinberg Equilibrium. To calculate the allele frequencies in these adults, which are also the allele frequencies in the gametes that will start the next generation, use the formula for the allele frequencies: Freq(A)=Freq(AA) + (1/2)Freq(Aa.) Once you have the allele frequencies that start the next generation, proceed as you would in the first case where you were given the allele frequencies directly. When e-mailing me problems, show your work as clearly as you can. You can use the following ways of representing things to make typing the formulas (which is tedious, as I know, just having typed out a bunch of them for this lecture) easier: To represent p squared (p2) type p^2. To represent w bar, the average population fitness, type wbar.

To represent things with subscripts, like the absolute and relative fitnesses, you don't have to put the subscripts as subscripts--it's OK to type:

It will probably help to work stuff with pen and paper first, then type it out. Get into the habit of writing down all formulas and steps clearly and legibly--you'll make many fewer mistakes, you'll be able to go back and study from what you've done, and I'll expect you to write down all formulas and steps clearly and legibly on exams so you'll do better if you get into the habit of doing so now. Finally, from time to time look at the questions and the answers and think about what you'd really expect to happen in the situation you're given, through natural selection. Do the numbers you're getting makes sense? Can you see how they might relate to a real situation? The goal, really, is to be able to translate evolution into math and then the math back into evolution--don't get so bogged down in algebra that you forget what you're really trying to study.

Now that you've learned the basic model, let's consider some more aspects of natural selection as it affects situations such as those you have modeled: natural selection occurring because of different fitnesses of genotypes for a single gene with two alternate alleles in a diploid, sexually reproducing population. Depending on the traits being studied, and the environment, fitness relationships among genotypes vary. They also depend on just how the traits are coded--whether one allele is completely dominant to the other (recessive) allele or whether there is some form of codominance or incomplete dominance such that the heterozygote has a different phenotype from either homozygote. (If these terms--dominance, recessiveness, incomplete dominance, codominance--are not completely familiar to you, be sure to review some genetics before proceeding!)

An important point to remember is that fitness depends on the phenotype--the appearance, structure, function of an organism. If two different genotypes result in the same phenotype (as when one allele is dominant) then both of those genotypes will have the same fitness--they look, act, function just like each other, so they survive and reproduce just like each other. Here are the possible ways the fitnesses of different genotypes could be related to each other: If one allele is completely dominant to the other, recessive, allele, so that the heterozygote has same phenotype as the dominant homozygote, there are two ways fitnesses could be related:

1. The recessive phenotype could have highest fitness (it would have a relative fitness value of 1; the dominant homozygote and heterozygote would have fitness values that are equal to each other and that are less than 1.)
2. The dominant phenotype could have highest fitness (in this case, the dominant homozygote and the heterozygote would both have relative fitness values equal to one, and the recessive homozygote would have a relative fitness value equal to something less than 1.)

If neither allele is completely dominant (in a case of codominance or incomplete dominance), so that the heterozygote has different phenotype from either homozygote, there are three ways that fitnesses could be related to each other:

1. Fitness codominance occurs when the heterozygote has fitness that is intermediate between the fitnesses of the two homozygotes (one homozygote has highest fitness, the heterozygote has lower fitness than it, and the other homozygote has still lower fitness.)
2. Heterosis occurs when the heterozygote has higher fitness than either homozygote (the heterozygote would have a relative fitness value of one, the two homozygotes would have fitness values of less than one.)
3. Underdominance (negative heterosis) occurs when the heterozygote has lower fitness than does either homozygote (one homozygote would have a relative fitness value of one, the other homozygote might also have a fitness value of one or, more likely, would have a somewhat lower fitness value, and the fitness value of the heterozygote would be even lower.)

You can use the formulas you've learned to calculate the effect of natural selection over one or two generations for any of these situations. If you repeated calculation of these formulas for 50 or 100 generations (I don't recommend doing this by hand!) you could learn what happens in each of the situations, in the long run. It turns out that the different situations listed above do differ in the expected result of natural selection. To learn about these, you will complete a computer exercise to study the different situations. In this assignment, you will look at specific instances of the forms of natural selection and of one other form of evolution, genetic drift. These specific instances illustrate general patterns--for example, what you learn for the specific case of heterosis you study will show you patterns that are generally true for heterosis. Your goal will be to learn these general patterns.

Embryogenesis Fertilizers

Embryogenesis begins when a sperm fertilizes an egg and creates a single cell that has the potential to form an entire organism. In the first hours after fertilization, this cell divides into identical cells. Approximately 4 days after fertilization and after several cycles of cell division, these cells begin to specialize, forming a hollow sphere of cells, called a blastocyst. The blastocyst has an outer layer of cells, and inside this hollow sphere, there is a cluster of cells called the inner cell mass. The cells of the inner cell mass will go on to form virtually all of the tissues of the human body.

Although the cells of the inner cell mass can form virtually every type of cell found in the human body, they cannot form an organism. Therefore, these cells are referred to as pluripotent, that is, they can give rise to many types of cells but not a whole organism. Pluripotent stem cells undergo further specialization into stem cells that are committed to give rise to cells that have a particular function. Examples include blood stem cells that give rise to red blood cells, white blood cells, and platelets, and skin stem cells that give rise to the various types of skin cells. These more specialized stem cells are called multipotent—capable of giving rise to several kinds of cells, tissues, or structures.

Cleavage Fertilization

After fertilization, the zygote proceeds immediately to the first cleavage (cell division) and subsequent cell divisions follow rapidly. The zygote is a very large cell, but the first waves of rapid cell division occur without increase in cell volume. The result is a closely bound mass of cells each of more typical cell size. At this stage the cells are called blastomeres, and the organism as a whole is called a morula from the time it has 16 blastomeres to the next stage. As it is undergoing this very rapid cell division, the organism is also migrating down the uterine tube towards the uterus.

Vertebrates, such as humans, have bilateral symmetry and therefore polarity in three dimensions (head-tail, or back-front, and left-right). Establishing polarity is one of the most basic manifestations of emerging specialization. But the egg is roughly spherical, and it is not readily apparent how polarity is established. Although it had been shown long ago that the point of sperm entry determines the plane of first cleavage (and thus subsequent ones) in amphibian eggs, mammals were believed until recently to remain spherically symmetrical until later in development. Recent data on mammalian zygotes, however, suggests that the point of sperm entry may similarly determine the cleavage plane. Even the first two cells resulting from the first cleavage may have different propensities, which persist through the next divisions as the progeny of one cell tend to become the body of the offspring and progeny of the other cell become the embryo's contribution to the placenta and other supporting structures. The word "fate," however, might be too strong, because the cells of such very early embryos are resilient to perturbations—if one cell is removed, the remaining ones can compensate.

Gastrulation Accumulate

Gastrulation begins when fluid starts to accumulate between the blastomeres of the morula. The fluid-filled spaces run together, forming a relatively large fluid-filled cavity. At the point when the cavity becomes recognizable, the organism is called a blastocyst. The outer cells of the blastocyst, especially those around the blastocyst cavity, assume a flattened shape. The flattened cells of the exterior blastocyst are the trophoblast. They become the embryo's contribution to the placenta and other supporting structures. On one side of the blastocyst is a group of cells that project inside into the blastocyst cavity; this is the inner cell mass, or embryoblast, and its progeny form the body of the new offspring.

The cells of the inner cell mass can give rise to progeny differentiating into all the types of cells in the adult body, so they are called pluripotent. They have not usually been described as totipotent because, the inner cell mass having already differentiated from trophoblast, the cells of the inner cell mass were believed to be no longer able to give rise to the cells of the trophoblast. Recent work, however, describes culture conditions under which human embryonic stem cells can differentiate to trophoblast cells. Although the new offspring itself develops only from the inner call mass, the trophoblast is not just passive padding. Its progeny are the essential and specialized connection between the embryonic and maternal systems. Embryonic stem cells can be isolated from the inner cell mass.

Organogenesis Structures

The basic structures and relations of all the major organ systems of the body emerge during the fourth through the eighth weeks of embryonic development. First, the embryo folds in several ways so that the flat linear structure distinguished by neural tube flanked by somites become roughly C-shaped. The effect of this is to bring the regions of the brain, gut, and other internal organs into their familiar anatomical relations. During the fourth week the neural pores, the ends of the neural tube "zipper," close. First the one at the cranial or head end, which is called the anterior or rostral pore, closes, and later the caudal or tail-ward pore closes.

Closure of the neural pores completes the closure of what will become the central nervous system. Also during the fourth week, limb buds become visible, first buds for arms and later for legs. Further, two accumulations of cells along the neural tube become distinguishable: the alar plate and the basal plate. Cells of the alar plate go on to become mostly sensory neurons, while basal plate cells give rise mostly to motor neurons. Already while the neural tube is closing, its walls along the cranial area are thickening to form early brain structure. Cranial nerves, for example the nerves for the eye and for the muscles of the face and jaw, also are beginning to develop at this time. The embryonic brain develops rapidly in both size and structure especially during the fifth week, and the optic cup that will form the retina of the eye becomes visible as well.

Homology Organisms

Similarities that exist between different organisms are cited by evolutionists as one of the most compelling lines of evidence for their theory. *Homologous* features are shared design concepts that are generally assumed by evolutionists to be the result of a shared ancestry and therefore offering support for the Darwinian theory of common descent. In The Origin of Species, Darwin referred to the *similar bones in the same relative positions* in the hand of a man and several other mammals as *specifically what the theory of common descent would expect.*

The 2008 edition of Prentice Hall Biology by Miller and Levine illustrates that this concept is commonly held wherein it states: *In their early stages of development, chickens, turtles, and rats look similar, providing evidence that they shared a common ancestry.*

Much like gene comparisons, the similarities that are evident during embryo development (*von Baer's law*) are assumed outright to be due to phylogenetic relationships and therefore homologous. Furthermore, embryo homology is considered to be such powerful evidence of common ancestry that patterns during embryogenesis are used to determine if shared morphological structures are truly homologous (due to shared ancestry) or merely analogous (due to convergent evolution). Miller and Levine define homology in such terms: *structures that have different mature forms in different organisms but develop from the same embryonic tissues.*

The reemerging of the early 19th century *"Von Baer's law"* has stimulated a new biological discipline called Evolutionary Developmental Biology, which is a collaboration of Evolutionary Biology and Developmental Biology, popularly known as Evo-devo. This area of scientific inquiry draw from the beliefs of Darwinism that all organisms have descended from the same ancestor, and generally assumes that embryonic developmental mechanisms should be shared (conserved). Drawing from these assumptions, evolutionary developmental biologists compare the developmental processes of different animals and plants in an attempt to determine the ancestral relationship between organisms.

Developmental biologist William Ballard a well known elder developmental biologist at Dartmouth College comments on the impact of *von Baer's law* on modern embryology.

engineered vaccines are also being developed to protect fish against pathogens. Genetic immunisation of rainbow trout with a glycoprotein gene from the virus causing viral haemorrhagic septicaemia has recently been shown to induce high levels of protection against the virus. Work is also underway on immunising carp, salmon and other fishes with genetically engineered vaccines for other diseases. The new molecular techniques are extremely sensitive and can identify pathogens in fish long before there are any clinical signs of the disease. This has implications for quarantine and the trade of aquatic species, which is currently governed by the World Trade Organisation and the Office International des Epizooties. Trade can be restricted based on the disease status of a product or a region; identification of minute quantities of a pathogen or of a new strain of an existing pathogen could change or influence existing trade patterns.

Farming Systems

Farming systems for aquatic species are diverse and include industrial scale farms, family ponds and culture based fisheries (stocking), in both developed and developing countries. Often, there is a division of the production process where fingerlings (*i.e.* small fish, especially up to one year of age) or eggs are produced by the seed-supplier, but the grow-out to market size is done elsewhere.

In the case of sea going salmon, there is often a seed supplier operating a hatchery near a river, a fingerling producer in a freshwater lake, and another group that grows the fish to market size in the sea. Marine shrimp hatcheries in Asia are usually small family owned ventures, whereas in Latin America they are more industrial in scale. Appropriateness of genetic biotechnologies must take these different systems into consideration.

Biotechnologies in Conservation

Genetic biotechnologies can be used to reduce the impacts of farmed fish on wild populations, to identify and manage endangered species and to manage captive populations in aquaria or in species recovery programmes. In several areas, farmed fish must be made triploid, *i.e.* sterile, in order to reduce their impact on wild populations should they escape from the fish farm. Generally, the planned use of transgenic fish also includes the provision that they are sterile, to reduce the chance of mixing with other fishes. Genetic manipulation and polyploidisation can be combined to regenerate endangered species. This can be done from frozen sperm by denaturing the DNA in an egg of a related species, fertilising with frozen sperm from the endangered species and then duplicating the chromosome-set of the fertilised egg.

Bibliography

Abhilash Jain and VarunKhetrapal.:*Environmental Biotechnology,* Pacific Books International, Delhi, 2012.

Arvind Kumar.:*Environmental Biotechnology,* Daya Publication, Delhi, 2004.

B.C. Bhattacharyya and Rintu Banerjee.:*Environmental Biotechnology,* Oxford University Press, Delhi, 2007.

Bimla Singh.:*Environmental Biotechnology,* Vista International Publication, Delhi, 2006.

C S K Mishra.:*Environmental Biotechnology,* APH Publication, Delhi, 2007.

Dilip Kumar Markandey and NeelimaRajvaidya.:*Environmental Biotechnology,* APH Publication, Delhi, 2004.

Geetha Bali, RallapalliRamamurthi, S.B. Sullia, Aziz Shiralipour and SatishKastury.:*Environmental Biotechnology,* APH Publication, Delhi, 2002.

H.P. Singh, Anil Kumar, V.A. Parthasarathy and Babita Singh.:*Advances in Horticulture Biotechnology Nanotechnology in Agriculture: Vol. 6,* Westville Publishing House, Delhi, 2013.

H.P. Singh, M. Anandaraj and A.I. Bhat.:*Advances in Horticulture Biotechnology Diagnostics for Horticulture Crops: Vol. VII,* Westville Publishing House, Delhi, 2013.

H.P. Singh, V.A. Parthasarathy and K. NirmalBabu.:*Advances in Horticulture Biotechnology : Molecular Markers and Marker Assisted Selection, Vol. III. Fruit Crops, Plantation Crops and Spices,* Westville Publication, Delhi, 2011.

H.P. Singh, V.A. Parthasarathy and K. NirmalBabu.:*Advances in Horticulture Biotechnology : Molecular Markers and Marker Assisted Selection, Vol. IV. Vegetables, Ornamentals and Tuber Crops,* Westville Publication, Delhi, 2011.

H.P. Singh, V.A. Parthasarathy and K. NirmalBabu.:*Advances in Horticulture Biotechnology : Regeneration Systems, Vol. I. Fruit Crops, Plantation Crops and Spices,* Westville Publication, Delhi, 2011.

H.P. Singh, V.A. Parthasarathy and K. NirmalBabu.:*Advances in Horticulture Biotechnology : Regeneration Systems, Vol. II. Vegetables, Ornamentals and Tuber Crops,* Westville Publication, Delhi, 2011.

H.P. Singh, V.A. Parthasarathy and K. NirmalBabu.:*Advances in Horticulture Biotechnology : Volume 5. Gene Cloning and Transgenic*, Westville Publication, Delhi, 2011.

HemantRawat.:*Environmental Biotechnology*, Oxford Book Company, Delhi, 2008 .

Mihir Kumar Das.:*Environmental Biotechnology and Biodiversity Conservation*, Daya Publication, Delhi, 2008.

P R Gilbert.:*Advances in Biotechnology*, Anmol Publication, Delhi, 2008.

P.R. Yadav and Rajiv Tyagi.:*Environmental Biotechnology*, Discovery Publishing House, Delhi, 2006.

Praveen K. Sharan.:*Advanced Biotechnology*, Campus Books, Delhi, 2010.

PurnimaSethi and V.S. Kulkarni.:*Environmental Biotechnology*, Alfa Publications, Delhi, 2011.

R A Sharma.:*Environmental Biotechnology*, Pointer Publication, Jaipur, 2007.

Rajiv K Sinha and RohitSinha.:*Environmental Biotechnology : Role of Plants, Microbes and Earthworms in Environmental Management and Sustainable Development*, Aavishkar Publication, Delhi, 2008.

Rajmohan Joshi.:*Environmental Biotechnology*, Isha Books, Delhi, 2006.

RamanandSagar.:*Advances in Agricultural Biotechnology*, Cyber Tech Publication, New Delhi, 2009.

Reza Marandi and Ali Shaeri.:*Environmental Biotechnology*, SBS Publication, Delhi, 2007.

S.K. Agarwal.:*Advanced Environmental Biotechnology*, APH Publication, Delhi, 2005.

S.K. Agarwal.:*Environmental Biotechnology*, APH Publication, Delhi, 2003.

S.K. Agarwal.:*Environmental Biotechnology*, APH Publication, Delhi, 2009.

S.N. Mukhopadhyay.:*Advanced Process Biotechnology*, Viva Books, Delhi, 2008.

Sanjay Harke, B.N. Pande and A.D. Diwan.:*Environmental Biotechnology and Sustainable Biodiversity*, Narendra Publishing House, Delhi, 2010.

Sohan Sharma.:*Advances in Fungal Biotechnology*, Random Publication, Delhi, 2012.

SoumitroGhose and S K Dubey.:*Environmental Toxicology and Biotechnology*, Dominant Publication, Delhi, 2009.

V.K. Saran.:*Environmental Biotechnology : Commercial Applications*, Medhashri Publications, Delhi, 2011.

V.S. Harikumar.:*Advances in Agricultural Biotechnology*, Regency Publication, Delhi, 2003.

Index

A

Accumulate 95
Administrative 238, 239
Aquaculture 235, 236, 237, 239, 240, 241
Aquaculture management 196
Aquaculture systems 28, 36
Aquatic products 169
Aquatic resources 161, 162
Artificial fertilization 78, 84, 128, 129
Assortment 185
Autosomes 187
Availability 237, 238, 240

B

Biochemical genetic markers 133
Biotechnology Research 104
Bodies 237, 238
Brackish 220
Breeding 240, 242

C

Cancer Cytogenetics 141, 156
Capture fishery 171, 200
Catfish 208, 213, 218, 219, 226, 227
Chondrichthyan fishes 79, 83
Class Petromyzontida 61, 67
Cleavage Fertilization 268
Collection 240, 242
Combination 218, 226
Communication 239, 241
Construction 237, 238
Conventional 237
Coral Reefs 247
Criteria 239
Culture systems 21

D

Derelict 238, 239
Designing 236
Digestive tract 272
Drainable 236, 237, 240
Drainage 210

E

Ecosystem 218
Efficient 239
Embryogenesis Fertilizers 267
Employment 236
Esterase 50
Exogenous 49
Extension 235, 236, 237, 238, 239, 240, 241

F

Farming 235, 236, 237, 238, 239
Fecundity 94, 96, 97, 98, 99, 188
Fertilization 190
Fisheries Technology 166, 201
Fishing Vessels 2
Food Industry 108
Freshwater fish 35, 170, 172, 195

G

Gastrulation Accumulate 269
Gene Amplification 158
Genetic Resources 243
Genetically discrete groups 134, 142
Genotype 96, 185, 186, 187, 188
Gynogenesis 184

H

Haploid 184, 185, 190
Harvesting 212, 224, 228
Homology Organisms 270
Homozygosity 91, 92, 95, 96, 97

I

Ictalurus 48
Implementation 240, 241
Important 237, 239, 241
Improper 238

"Thus, the energy of investigators and particularly students is diverted into the essentially fruitless 19th century activity of bending the facts of nature to support second-rate generalities of no predictive value. Though enthusiasm for Haeckel's recapitulation `law' died out, unfortunately the popularity of Von Baer's `laws' of 1828 was renewed. In order to defend the latter's descriptive statements that general characters appear before special characters as an egg develops and that the less general and finally the specific characters trail along later, we have to decide intuitively that certain characters are of `morphological significance' and others are not. When referring to vertebrates, we have to use words like blastula and gastrula in such a way as to imply that things that are vastly different from each other are really very much the same. "

Differences among organisms have traditionally been assumed to become more significant as development proceeds, and therefore earlier embryonic stages should resemble one another more than latter stages. This view can be found presented in popular biology textbooks. Although these assumptions have persisted into modern times, it has been recognized by embryologists since Darwin that this doctrine of developmental biology *"is not in accordance with the facts of development"*, according to British embryologist Adam Sedgwick who also stated: *there is no stage of development in which the unaided eye would fail to distinguish between them with ease"*.

Modern embryologists confirm this as well, and have further challenged the erroneous nature of these assumptions. Recent work has revealed that embryos are not more similar during the earlier stages of cleavage and gastrulation, but in fact, are most alike at the later stage illustrated by Haeckel and recently photographed by Richardson. In the gastrulation stage, fish show dramatic differences when compared to amphibians, and neither are similar to reptiles, birds, or mammals. Organisms begin life very dissimilar, then become somewhat alike during embryo development, then diverge again as adults. Evo-devos (evolutionary developmental biologists) refer to this pattern as the *developmental hourglass*. William Ballard adds that *only by semantic tricks and subjective selection of evidence can we claim that "gastrulas" of shark, salmon, frog, and bird are more alike than their adults.*

Ballard continues: "In fact, the most obvious structural characteristics of either the eggs or the cleavage stages of a shark, a salmon, a frog, a bird, or a mammal are unique each to its own class, not generally shared. We would not consider them very much alike unless we had been taught so at a very early age. Very few vertebrates pass through a stage which can strictly be called a blastula. The embryo in its period of most active morphogenetic movements is usually called a gastrula, but as all agree this word has no morphologic meaning anymore. Each class of vertebrates (in mammals we might almost say each particular order) develops and then loses its own set of temporary structures-like the parade ground `formations of maneuver' during this period. "

Another significant challenge for evo-devo is the recognition that organs and structures develop from different groups of embryo cells in organisms that are believed to share a common ancestry. Fundamental structures, like the alimentary canal, kidney, or forelimb are commonly found to form from different embryological tissues. The alimentary canal (digestive tract) is formed from the roof of the embryonic gut cavity in sharks; whereas in frogs it is formed from the gut roof and floor; and in birds and reptiles it is derived from the lower layer of the embryonic disc or blastoderm. Vertebrate forelimbs, which were used as an example of homology by Darwin and can be found in hundreds of textbooks as proof for evolution, also develop from different body segments in different species. Rather than being proof of Darwinism, recent embryological studies have shown that forelimbs are not developmentally homologous. The forelimbs in the newt develop from embryo trunk segments 2 through 5; in the lizard they develop from segments 6 to 9; in humans they develop from segments 13 through 18. The development of the vertebrate kidney also conflicts with the view of homology. Evolutionist Michael Denton notes these conflicts in Evolution: A Theory in Crisis.

"In fish and amphibia the kidney is derived directly from an embryonic organ known as the mesonephros, while in reptiles and mammals the mesonephros degenerates towards the end of embryonic life and plays no role in the formation of the adult kidney, which is formed instead from a discrete spherical mass of mesodermal tissue, the metanephros, which develops quite independently from the mesonephros.

Studies of the genes governing development have added even further difficulties for evo-devo. According to the views of neodarwinists, homologous features are produced by genes that have been passed down by common ancestors. Because similar genes are unlikely to have originated independently through random mutation, sequence similarity between organisms indicates they share an ancestral relationship (they are homologous). Features produced by similar sequences are then also inferred to be phylogenetically homologous. For decades it has been recognized that homologous organs were frequently produced by different genes, but more recently, evo-devo biologists discovered that certain developmental genes were highly conserved (similar) in radically different organisms. These homeotic genes are short sequences that act as regulatory switches affecting the development of structures such as eyes and legs. When mutated, homeotic genes can transform one body part into another, such as an antenna into a leg. In organisms as vastly different as mammals and fruit flies, these genes are so similar that the mouse gene can regulate the development of the compound eye of the fruit fly. In such examples, the conflict with Darwinian assumptions lies in the fact that non-homologous structures are produced (or at least regulated) by supposedly homologous genes. As Jonathan Wells states very poignantly: *homology, whether defined morphologically or phylogenetically, cannot be attributed to similar developmental pathways any more than it can be attributed to similar genes.*

The famous embryologist Sir Gavin de Beer challenged his colleagues with the following question which remains unanswered to this day.

"But if it is true that through the genetic code, genes code for enzymes that synthesize proteins which are responsible (in a manner still unknown in embryology) for the differentiation of the various parts in their normal manner, what mechanism can it be that results in the production of homologous organs, the same "patterns", in spite of their not being controlled by the same genes? I asked this question in 1938, and it has not been answered.

Worldviews

While Darwin was convinced that embryological homology was the *strongest single class of facts* to support his views. Modern embryology has instead challenged the Darwinian theory of common descent by illustrating that seemingly homologous structures, such as the mammalian forelimb, developed along different pathways, and are therefore not homologous at all. While the Darwinists see these similarities as evidence of common descent, the creationist belief system causes us to generally conclude that shared structures, processes, or genes are merely reflections that they were designed by the same creator. Rather than being evidence of shared ancestry, homologies are examples of brilliant and well functioning designs that have been applied to multiple organisms, much in the same way as human designers apply concepts.

Biotechnologies in Aquaculture

Genetic biotechnologies in aquaculture focus primarily on increasing growth rate, but also include disease resistance and increased environmental tolerance. There are several biotechnologies that can be applied to farmed aquatic species.

Selective breeding, *i.e.* traditional animal breeding, started with the common carp several thousand years ago. However, it has only recently been applied to a handful of other species of food fish such as catfish, trout and tilapia. Therefore, many farmed aquatic species are very similar to their wild relatives. Selective breeding programmes have yielded significant and consistent gains of 5-20% per generation in species of, inter alia, Atlantic salmon, catfish and tilapia. Hybridisation is a simple genetic technology that has become easier with the development of artificial breeding techniques, such as the use of pituitary gland extract and other hormones to initiate gamete development and induce spawning (*i.e.* the depositing of eggs), and an increased understanding of environmental cues that influence reproduction, such as day length, temperature or water current. Many of the natural reproductive isolating mechanisms that species develop in the wild can now be overcome by fish farmers.

These improvements in reproductive technologies have also assisted aquaculturists greatly in their efforts to domesticate aquatic species. In addition, by making it possible to remove the natural constraints and timing of breeding, farmers are able to mate many more species at the times that are most beneficial, and thus help to ensure a steady and consistent supply of fish to the market. Chromosome-set manipulation can be used to produce triploid organisms that generally do not channel energy into reproduction because of problems associated with development of reproductive organs. Initially it was thought that this energy saving would result in increased growth rate, but this seems not to be the case. The real advantage of triploids seems to be in their functional sterility.

For example, triploid oysters do not produce gonads (*i.e.* reproductive glands) and are therefore marketable at times of the year when mature oysters have an off-taste because of production of gametes (*i.e.* sex cells - the ovum, or egg (female), and sperm (male)). In aquaculture, one sex is often more desirable than the other. For example, female sturgeon produce caviar, male tilapia grow faster than females whereas it is the female trout and salmon that generally grow faster than the males. The production of single sex groups of fish takes advantage of these differences between the sexes and can be accomplished by manipulation of the developing gametes and embryo.

The manipulation can be in the form of denaturing (*i.e.* destroying) the DNA in gametes followed by chromosome-set manipulation or by hormonal sex-reversal and subsequent breeding. The phenotypic sex of many aquatic species can be changed by administering appropriate hormones.

For example, genetically male tilapia can be turned into females through estrogen treatments. These genetic males when mated with normal males produce a group of all-male tilapia that grow faster and have less unwanted matings (that lead to overcrowding and stunting) than a group of mixed-sex tilapia. Some of the all-male offspring would have two male chromosomes and these could be used as broodstock for subsequent generations, thus avoiding the use of hormones in the broodstock.

Hybridisation can also be used to produce single sex groups of fish, when the sex-determining mechanisms in the parental lines are different (for example, hybridisation of Nile tilapia and the blue tilapia). Genetic engineering is a vague term that has come to be nearly synonymous with gene transfer *i.e.* the production of transgenic fish or genetically modified organisms (GMOs). This technology is progressing rapidly and it is now possible to move genes between distantly related species. Gene transfer in fish has usually involved genes that produce growth hormone and has been shown to dramatically increase growth rate in carp, catfish, salmon, tilapia, mudloach and trout. A gene from the winter flounder that produces an anti-freeze protein was put into salmon in the hope of extending the farming range of the fish.

The gene did not produce enough of the protein to extend the salmon's range into colder waters, but it did allow the salmon to continue growing during cold months when non-transgenic salmon would not grow. Transgenic technology is currently in the research and development stage. To our knowledge there are no transgenic aquatic plants or animals available to the consumer.

Cryopreservation

The development of cryopreservation or low-temperature technology allows the short- and long-term storage of gametes. Currently, these low-temperature techniques can only be used on male gametes; eggs and embryos can generally not be stored in this way. Freezing gametes can increase the flexibility of a fish breeder, especially when breeding species where the sexes mature or migrate at different times, when the breeding season is very short, when the breeders are far apart or when one sex is exceptionally rare.

Fish Health

Genetic biotechnologies are being used to improve fish health through conventional selection for disease resistance and through the use of molecular investigation of pathogens for characterisation and diagnosis.

DNA-based technologies are being used now to characterise different species and strains of pathogens. Genetic characterisation of the pathogen may also reveal information about its origin, *e.g.* DNA analysis revealed two strains of crayfish plague fungus in Sweden: one from the local species and one originating in Turkey. Once the pathogen is characterised, DNA probes can be developed to screen for specific pathogens in tissue, whole animals and even in water and soil samples. These techniques are being used to detect viral diseases of marine shrimp throughout the world and for bacterial and fungal pathogens in fishes in many areas. Genetically

Inorganic fertilizers 21
Interspecific hybridisation 175
Intrinsic 42
Inventory 237, 239
Irrigation 210

J

Juvenile 208
Juveniles 168

L

Laboratory techniques 153
Lampreys 57, 58, 61, 67
Lipids 44, 46, 47, 50, 51
Liver 241

M

Major 237, 238, 240
Manipulation 184
Manure 216, 217
Marine biotechnology 166, 202, 258
Marine ecosystems 248
Marine Fisheries 16
Meroblastic pattern appears 77, 78
Microdeletion Syndromes 139
Monoculture 212, 213, 226, 227
Monomorphic 185

N

Natural collections 78
Nervous system 130, 133, 135, 143, 269
Nursery 212, 226, 227, 240
Nutrition 236

O

Oesophagus 40, 41, 42, 47, 48, 50
Operation 235, 236, 239, 240
Organogenesis Structures 269
Ownership 220

P

Peristalsis 42, 47
Plant biotechnology 102, 104
Pond aquaculture 20
Pond construction 20
Preparation 241
Progeny 93, 186

Q

Quality 238, 240, 241

R

Ratcheting 40
Renovation 238
Reservoir fisheries 195
Riverine 238

S

Spermatogenesis 83
Springs 214
Stocking 222
Supplementary 238
Survey 236, 237, 239
Sustainable development 32
Sustainable fisheries 1
Sustainable Production 244

T

The Skeleton 52
The skeleton 52
Topography 209, 212
Toxicants 238
Transport 239
Triturating 40
Tropical fish 118

U

Unconventional 237
Undrainable 236, 237, 240

V

Vaccines Development 196
Village 236, 237, 238, 239, 242